UNDERDETERMINATION

BOSTON STUDIES IN THE PHILOSOPHY OF SCIENCE

Editors

ROBERT S. COHEN, *Boston University*
JÜRGEN RENN, *Max Planck Institute for the History of Science*
KOSTAS GAVROGLU, *University of Athens*

Editorial Advisory Board

THOMAS F. GLICK, *Boston University*
ADOLF GRÜNBAUM, *University of Pittsburgh*
SYLVAN S. SCHWEBER, *Brandeis University*
JOHN J. STACHEL, *Boston University*
MARX W. WARTOFSKY†, (*Editor 1960–1997*)

VOLUME 261

UNDERDETERMINATION

An Essay on Evidence and the Limits of Natural Knowledge

by

THOMAS BONK
*Ludwig-Maximilians-Universität
München, Germany*

Library of Congress Control Number: 2008921235

ISBN 978-1-4020-6898-0 (HB)
ISBN 978-1-4020-6899-7 (e-book)

Published by Springer,
P.O. Box 17, 3300 AA Dordrecht, The Netherlands.

www.springer.com

Printed on acid-free paper

All Rights Reserved
© 2008 Springer Science + Business Media B.V.
No part of this work may be reproduced, stored in a retrieval system, or transmitted
in any form or by any means, electronic, mechanical, photocopying, microfilming, recording
or otherwise, without written permission from the Publisher, with the exception
of any material supplied specifically for the purpose of being entered
and executed on a computer system, for exclusive use by the purchaser of the work.

Preface

Any set of phenomena admits of more than one theoretical explanation. In the short or long run testing and experimentation and basic methodological maxims tend to reduce a set of competing explanations to one. But occasionally this trusted procedure appears to fail spectacularly: competing theories are observationally equivalent, matching prediction by prediction. A well-tested explanation of phenomena turns out to have an observationally equivalent theoretical rival. Since collecting more data appears to be useless, and simplicity is not an infallible guide, one is left to wonder: which of the rival theories ought one to believe, or should one suspend judgement perhaps indefinitely? The 'underdetermination' of scientific theories by the data they explain has been given wider epistemological and ontological significance. W. V. Quine, in particular, suspected an "omnipresent under-determination of natural knowledge generally". Suggestive examples from physics, methodological and semantic arguments all have been advanced to show that underdetermination runs deep and pervades our knowledge of the world. If rival theories can meet ideal methodological requirements for justified belief in equal degree, and still differ radically in what they claim about reality, is it not better then to adjust our ideas about reality, reject scientific realism, and radically revise our concept of justification and knowledge?

These ideas have met with stiff resistance and increasing skepticism. Many philosophers of science today find the examples provided too limited and the systematic arguments somehow unconvincing. The preferred approach by scientific realists is a methodological one. They hold that proponents of a general underdetermination of theories by data are in the grip of a false and unrealistic methodology of testing and confirmation. Some have advanced novel concepts of support and methodological rules to remedy the situation. The second main response to the 'omnipresent under-determination of natural knowledge' has been somewhat neglected in comparison. It aims to resolve underdetermination issues by reflection on the way we conceptually represent the facts to us. The most well-known approach in this family is conventionalism, which offers a diagnosis for why underdetermination arises and an elegant solution to the epistemological problems underdetermination seems to create as well. Quine's own 'pragmatic' interpretation of under-determination belongs to this broad family.

Analyzing and evaluating the nature of underdetermination of theory by observation in full is a tall order. The project touches on many philosophical subjects and would take a much longer and more comprehensive work than this one. The goal of my essay then is *propaedeutic*: to clarify the question, to analyze and balance key reasons for and against believing that significant 'empirically irresolvable' conflicts exist, and to examine how far major philosophical responses to the problem are cogent. Since the issue of underdetermination has been brought to the fore by W. V. Quine, I find it necessary and illuminating to discuss pertinent aspects of his views on the matter at hand.

From realist critics of the premise I differ on two points: I do not think that mutually observationally underdetermined theories in the sciences are occasional mishaps; and I have little expectation for strengthened accounts of justification to deliver theories uniquely. From pragmatists like Quine I differ in that I think that the 'no fact to the matter' response should be resisted except in a well-defined set of circumstances, and that considerable and persistent disagreement between rational investigators under ideal epistemic conditions is an intelligible possibility. What can be shown is that any inference from observational underdetermination to the non-existence or indeterminacy of an entity or concept is far from automatic. It depends on a variety of additional assumptions, which on examination turn out to be questionable if not outright question-begging. I want to deny, in particular, that there is necessity or inevitability in the existence of observational equivalent, alternative theories. This premise is crucial for arguing from the underdetermination thesis to an anti-realist conclusion. In this piecemeal way the anti-realist momentum, which observationally 'irresolvable conflicts' between theories frequently carry, is blocked.

The introductory chapter assesses broadly the metaphysical and epistemological significance of the underdetermination thesis. To this end I distinguish several kinds of underdetermination of theory by evidence of different strengths. I examine the conflict between the doctrine of scientific realism and the thesis of underdetermination by evidence, and identify major responses to the problem and their conceptual inter-relations.

The second chapter is a survey of the cornerstones of the philosophical debate that is the subject of my essay. I review basic logico-mathematical notions of equivalence and relevant semantic theorems. These are frequently referred to in an effort to bolster arguments for or (less frequently) against systematic underdetermination. The theorems of Craig and Ramsey have a special significance in this respect. I look at a few examples of underdetermined physical theories and observationally 'equivalent' physical systems, some well-known some less so.

The starting point of chapter 3, "Rationality, Method, and Evidence", is the claim that deductivism as a method of testing, confirming hypotheses and justifying belief in theories gives inevitably rise to 'empirically irresolvable' conflicts of assertion. Quine's radical epistemology seems to rely on deductivism and related doctrines, like holism and Duhem's thesis. Much attention by critics has been lavished on this point in the hope to show that by shoring up our standards of justification, and by giving a fuller and more realistic account of scientific methodology, all *prima facie* cases of observational underdetermination at all levels can be satisfactorily resolved. I re-examine deductivism and compare the limitations of three recent accounts of empirical support: instance confirmation, demonstrative induction, and evidentially relevant relations between independent theories. While hypothetico-deductive methodology is clearly flawed, these accounts of testing and confirmation, despite some promising results, offer no general solution to the problems posed by underdetermination at present, I conclude.

In chapter 4, "Competing Truths", I examine, first, ways of systematically generating observationally indistinguishable alternatives to any theory, however well supported by tests. Typically these are semantic arguments, which aim to show that theories with different and incompatible universes of discourse or 'ontologies', agree on all observational consequences. I count among them 'proxy-function' constructions, switch to coordinate languages, permutations and Löwenheim-Skolem type arguments. Without aiming to resolve all the methodological, logical, and epistemological issues involved in assessing these semantic arguments and "algorithms", I argue that arguments of this kind depend on questionable assumptions, which tend to erode their metaphysical and epistemological significance.

Chapter 5, "Problems of Representation", examines the neglected idea that the origin of the conflict lies entirely with our conceptual representation of the facts, i.e. in the way we represent the world in our theories. The best known and most enduring form is conventionalism. P. Horwich advanced the idea that perhaps all genuine cases of underdetermination are between "notational variants". D. Davidson suggested that changing the terminology may lift any conflict between mutually underdetermined rival theories or total science. I aim to show, by moving from simple to more sophisticated versions that these approaches to underdetermination, on the whole, are unsatisfactory.

My starting point in the final chapter 6, "Underdetermination and Indeterminacy", which continues the discussion of the approach of the previous chapter, is the well-known argument from the underdetermination of a translation scheme to the "indeterminacy of translation" thesis. It is an exemplary instance, I believe, of the reasoning leading to Quine's surprising claim about the limitations of natural knowledge, quoted at the beginning. I unravel some of the threads in Quine's influential argument, and examine Quine's eventual pragmatic interpretation of the underdetermination thesis. H. Field advocates the introduction of non-standard degrees of belief to reflect adequately frequent indeterminacy in our language. I argue that observational underdetermination of conflicting assertions is not in general a sufficient criterion for there being 'no fact to the matter', which of the assertions is correct.

This project has grown over the years in leaps and bounds. I have learned in conversations and lectures from more people than I can now remember. Those that stand out most prominently in my mind are Bas van Fraassen, Jesus Mosterin, C. Ulises Moulines, Sir Michael Redhead, Nicholas Rescher. I thank John D. Norton for conversations pertaining to the theory of confirmation and Fred Kronz for reading a very early draft of what has now become chapter 2. I began developing my ideas on observationally underdetermined theories during a stay at the Center for Philosophy of Science at the University of Pittsburgh on a generous *Alexander von Humboldt-Fellowship*, for which I am grateful.

Contents

1 A Humean Predicament?   **1**
   1.1 Aspects of Underdetermination   1
   1.2 Significance of the Thesis   8
   1.3 Quine, Realism, and Underdetermination 21
   1.4 No Quick Solutions 28
   1.5 Three Responses and Strategies 38

2 Underdetermination Issues in the Exact Sciences   **45**
   2.1 Logical Equivalence, Interdefinability,
        and Isomorphism 45
   2.2 Theorems of Ramsey and Craig 52
   2.3 From Denotational Vagueness to Ontological Relativity 58
   2.4 Semantic Arguments 61
   2.5 Physical Equivalence 68
   2.6 Underdetermination of Geometry 82

3 Rationality, Method, and Evidence   **89**
   3.1 Deductivism Revisited 89
   3.2 Quine on Method and Evidence 102
   3.3 Instance Confirmation and Bootstrapping 112
   3.4 Demonstrative Induction 119
   3.5 Underdetermination and Inter-theory Relations 126

4 Competing Truths   **141**
   4.1 Constructivism 141
   4.2 Things versus Numbers 147

4.3	Squares, Balls, Lines, and Points	161
4.4	Algorithms	167

5 Problems of Representation 177
5.1	Ambiguity	177
5.2	Conventionalism: Local	180
5.3	Conventionalism: Global	189
5.4	Verification and Fictionalism	196

6 Underdetermination and Indeterminacy 207
6.1	Underdetermination of Translation	207
6.2	Indeterminacy versus Underdetermination	215
6.3	Empirical Investigations of Cognitive Meaning	230
6.4	Indeterminacy and the Absence of Fact	238
6.5	Quine's Pragmatic Interpretation of Underdetermination	252

Bibliography 259

Index 279

Chapter 1
A Humean Predicament?

In this chapter I examine the significance of observationally 'irresolvable conflicts of assertion' and connect the problem with its rich history. I look at three prominent and plausible ways to diagnose the problem.

1.1 Aspects of Underdetermination

The idea that scientific theories are underdetermined by observations and experiments that appear to support them did not originate with Willard V. Quine, but he did much to bring out its epistemological, methodological and metaphysical significance. To have a basic formulation of ('strong') underdetermination of theories in the natural sciences at hand Quine's original formulation is a good starting point (he also used 'partially determined' and 'indeterminacy of determination' instead of underdetermination).

> Scientists invent hypotheses that talk of things beyond the reach of observation. The hypotheses are related to observation only by a kind of one-way implication; namely, the events we observe are what a belief in the hypotheses would have let us to expect. These observable consequences of the hypotheses do not, conversely imply the hypotheses. Surely there are alternative hypothetical substructures that would surface in the same observable ways. Such is the doctrine that natural science is empirically underdetermined; under-determined not just by past observation but by all observable events. (Quine, 1975, p. 313)

The concept of empirical underdetermination has many dimensions, and we will encounter variant formulations throughout the essay. The statement quoted indicates some of the *conceptual variables* of the underdetermination thesis that shaped the subsequent debate: the entities that are thus thought to be empirically underdetermined – 'natural science', 'systems of belief', theories, theories together with a belt of auxiliary hypotheses or 'total theories'; the account of justification, *scientific methodology* and reduction involved when talking about evidence and observation; the notion of *observational equivalence* and empirical content; and the character of the data and observation which are underdetermining theories. Quine's statement does not merely assert that no scientific theory worthy of that name can be conclusively 'verified' (its truth-value 'determined') no matter how many observations are taken into account, while the truth-value of, say, "all ravens are black" can be determined in the limit. Rather the claim is that theoretical terms have no reductive definitions in terms of the chosen observational vocabulary, and hypotheses essentially involving theoretical terms are not verifiable one by one. Theoretical terms are 'free inventions', only deductively grounded in the facts to be explained. Hence, it is likely that theoretical alternatives exist to any given mature and successful theory that reproduce all relevant observations of the latter (presumably without being 'rival' linguistic variants thereof).[1] In this sense "natural science is empirically underdetermined". There are many statements like the one above scattered throughout Quine's work, indicating that he takes the two-step process of theoretical "postulation" and deduction of observation sentences as fundamental (not only to the methodology of the sciences, but to epistemology as well). In strict deductivism, not truth but falsity is transmitted from observation sentences to the hypotheses: falsification rules; but Quine's deductivism is not of the strict kind as we shall see. More recently, R. Boyd expressed the underdetermination thesis in the following way:

> A theory is said to be empirically equivalent to T a pro-

---

[1]Strictly speaking 'alternatives' means exactly two options. Following common, albeit loose usage I will use 'alternatives' to cover any number of mutually incompatible, rival statements, theories, etc.

posed theory just in case it makes the same predictions about observable phenomena that T does. Now, it is always possible, given T, to construct arbitrarily many alternative theories that are empirically equivalent to T but which offer contradictory accounts of the nature of unobservable phenomena. (Boyd, 1984, p. 42f)

Here theories and scientific methodology are central, not 'global systems of belief' and questions of reduction and definability. The thesis of empirical underdetermination of theories has two aspects. First, mutually contradicting scientific theories may be fully empirically equivalent, have identical empirical content or make exactly the same predictions (Boyd). In other words, to any given theory (or theory with 'sufficient critical mass') there is an alternative, which has the same observational content. Although the crucial role of auxiliary hypotheses in deriving testable results from theories is not explicitly mentioned in the quotes, it is easy to reformulate the thesis of underdetermination to take this dependence of the observational content on auxiliaries into account.[2] Second, ideal knowledge of "all observable events", past, present and future, cannot eliminate the possibility that 'natural science' or 'global science', which ideally accounts for those observations, has contradicting alternatives (Quine). In other words, our scientific conception of the natural world, by its own lights, need not be unique in point of ontology or what the fundamental laws of the world are. One can move from the first to the second claim by substituting for 'theory' the more glib 'global science', but the emphasis is different and it is better to keep them separate. It is seems to be consistent to accept the second claim while denying the first one.

---

[2] I adopt for the purpose of my argument the "statement view" of theories since it is championed by the main protagonists of underdetermination discussed in this essay. Briefly, a 'theory' is a finite set of statements which contain non-observational predicates essentially and allows for the derivation of an infinite number of observable consequences. I bracket the eminent possibility that the orthodox explication of theory is to be replaced by a structuralist account like the one developed in Balzer et al. (1987). I have found no indication that adoption of the latter account helps very much with clarifying the underdetermination issue.

The crucial concept of empirical or observational equivalence can be understood in more than one way. One may think of matching predictions, of equivalence in terms of identical observational consequences (which are thought to exhaust all the relevant evidence), or of the existence of a mapping between empirical sub-structures of the models associated with alternative theories. I examine a narrower and in certain contexts perhaps more useful notion of "physical equivalence" in chapter 2; for some remarks concerning the related notions of 'observation' and 'possible observation' see section 3.2. On any account, theories that are logically equivalent are empirically equivalent, but not vice versa. I take up the looming implicit question of the proper relation of confirmation between theory and data, hence what can count as evidence, in chapter 3.

The claims above circumscribe what might be termed *strong* underdetermination of theories or systems of theories (mainly with a view to natural science) in contrast to *weak* underdetermination,[3] also called "*Humean*" underdetermination (L. Laudan). The latter refers to the familiar fact that more than one hypothesis fit a given, finite set of quantitative data generated by an experiment, observation or measurement. Most of those weakly underdetermined hypotheses interpolate the same set of data in different ways and make different predictions. New data will falsify many of the (known or unknown) hypotheses, but at each stage indefinitely many un-eliminated ones exist. The link between weak and strong underdetermination will be examined more closely below in section 1.5. A variant is "*methodological*" underdetermination of a hypothesis (D. Mayo), which refers to the situation when the hypothesis and its rival both pass the same *severe* tests. A severe test of a hypothesis is one, roughly, that it is highly unlikely to pass if the hypothesis tested were actually false. "*Transient*" underdetermination (L. Sklar) is another variant of weak underdetermination. It refers to a passing situation in which the scientific community is (temporarily) unable to clearly reject one and confirm the

---

[3] I borrow the terminology from C. Hoefer and A. Rosenberg, (1994), who use it to draw a different distinction between theories that agree in their predictions only in the actual world (weak), and those that agree even beyond the actual world (strong). I hope no confusion results.

other of two alternative theories on account of their respective fit with the data alone. Such a situation can arise from many causes, among them a dearth of data, the lacking quality of data, a lack of variety among relevant observations, or lacking mathematical knowhow for deriving novel, testable predictions. Until these causes are removed, and consequently respective predictions significantly differ, scientists make their choices apparently on other grounds, frequently by way of simplicity comparisons, and sometimes on extra-evidential grounds or guided by non-cognitive factors. An often cited example is the evidential "stand-off" between (various forms of) the Copernican and the Ptolemaic system of celestial motion between roughly the publication of *De Revolutionibus* in 1543 and the publication of Kepler's *Harmonice Mundi* in 1619 and beyond. Both astronomers became convinced of the falsity of the Ptolemaic system mainly by what they perceived as the greatly increased (parametric) simplicity of their respective theories, apart from the broader metaphysical views they held. The reader can consult Laudan (1996, pp. 34f) for further distinctions among underdetermination claims, like "deductive" underdetermination, "ampliative", or "descriptive" underdetermination. I do not mention them explicitly here, since they appear to reflect specific *reasons* advanced for believing in universal (strong) underdetermination of scientific theories, in particular differences of opinion over the proper methodology of testing and confirmation. I examine the methodological assumptions separately in a chapter below.

In Laudan (1996) and Leplin (1997b, pp. 204f) the authors introduce a helpful way of framing strong underdetermination claims. They reserve the label underdetermination ("UD") for the epistemological claim that no amount of evidence can warrant belief in any theory, in what the theory literally understood says about the unobservable. A further, independent thesis ("EE") says that any theory has incompatible rivals with exactly the same observational consequences, or the same empirical deductive content. The task is, as Laudan and Leplin see it, to investigate whether the latter thesis or a variant is true, and if true whether the inference from EE to UD is defensible. Helpful as this way of thinking through the issue may be, originally and more usually the label underdetermination is applied to claims like the empirical equivalence thesis EE and not to UD, as the quotes above indicate. On the other hand, Quine, the prime suspect, would not subscribe to UD, at least not in the bald way the claim is stated. Laudan and Leplin's proposal for framing the issue is but a way to emphasize (justly) the need to investigate the dependence of any claim of strong underdetermination on an account of evidence and scientific methodology. But one can do so directly, and I will not often take advantage of their way of structuring the issue.

Awareness of the apparent looseness of our grip on the unobservable in predicting the observable considerably pre-dates the authors quoted. Long before O. Neurath, P. Duhem, W. Whewell and J. S. Mill, to name a few, *Descartes* cautioned at the end of a comprehensive and detailed physical explanation of planetary motions, of comets and fix stars, gravity, magnetism, the behavior of the elements of earth, the air, the tides, earthquakes, fire, water and oil, and the functioning of our sense organs, based on the assumption that there are sub-visible particles:

> But here it may be said that although I have shown how all natural things can be formed, we have no right to conclude on this account that they were produced by these causes. For just as there may be two clocks made by the same workman, which though they indicate the time equally well and are externally in all respects similar, yet in nowise resemble one another in the composition of their wheels, so doubtless there is an infinity of different ways in which all things that we see could be formed by the great Artificer [without it being possible for the mind of man to be aware of which of these means he has chosen to employ]. (Haldane and Ross, 1955, p. 300)[4]

Descartes seems here to address the issue of strong underdetermination of global science. He offers a probabilistic argument to the effect that if one finds a theory, which accounts comprehensively for such a diverse

---

[4]Text in brackets inserted by the editors, an interpolation from the French version of the Latin text.

1.1. Aspects of Underdetermination

set of phenomena on the basis of a few general principles, at all than it is likely to have identified the true causes. Thus, he thinks (wrongly), strong underdetermination is defeated.

It is to the traditional criterion of *simplicity*, that Einstein and Leopold Infeld turn to after introducing their readership to the phenomenon of theoretical underdetermination: "Physical concepts are free creations of the human mind, and are not, however it may seem, uniquely determined by the external world. In our endeavor to understand reality we are somewhat like a man trying to understand the mechanism of a closed watch. [...] If he is ingenious he may form some picture of a mechanism which could be responsible for all the things he observes, but he may never be quite sure his picture is the only one which could explain his observations. He will never be able to compare his picture with the real mechanism and he cannot even imagine the possibility or the meaning of such a comparison" (Einstein and Infeld, 1938, p. 31). Einstein and Infeld take up the very same metaphor of nature as a closed watch (albeit dropping reference to the creator) that Descartes had used earlier. They went on to claim that the theoretical virtue of simplicity is usually the best criterion for discriminating between equivalent alternatives (Einstein and Infeld, 1938, pp. 31, 237). Interestingly, Einstein and Infeld do not take this problem to undermine in general a realistic interpretation of the sciences.

To return to Quine, *indeterminacy* is a related watchword. To claim that it is (objectively) 'indeterminate' whether x or y is to claim that x are y are mutually empirically underdetermined *and* that (therefore?) any debate whether it is the case that x or y is factually vacuous – there is no fact to the matter. Questions regarding indeterminate choices do not have "unknowable" answers, they have objectively no rational answer at all. The expression 'indeterminacy' in its many uses indicates then a potential, controversial *diagnosis* of certain cases of empirical underdetermination. Not all cases of underdetermination (particular in the sciences) point to an 'indeterminate' concept or state of affair as their source. The diagnosis 'indeterminacy' makes strong claims with regard to the maximality of the set of sentences (the evidence) on which the conflict of assertions is potentially to be decided, conditions that may or may not be satisfied. Typically this set will have to be the ideal totality of physical facts and perhaps observational laws, if (empirical) indeterminacy is to established. The appeal to this idealized totality of facts is both plausible and problematic, the latter on account of the observation–theory dichotomy and the role of counterfactuals. Although the *terminus technicus* indeterminacy in the present philosophical context is due to Quine, its content, significance and his motive in introducing the term would be recognizable instantaneously to H. Reichenbach. I discuss implications between the concepts underdetermination and indeterminacy, and other issues surrounding the positing of an 'indeterminacy' in chapter 6.

In the following I employ a slightly different terminology, both to generalize features of the problem we are facing and to steer clear of certain ingrained ways of conceiving the underdetermination of theories. I call alternative theories 'observationally indistinguishable' if there is no crucial observation or singular fact (known or unknown) that is consistent with (or confirms) one alternative and is inconsistent with the truth of the other (or disconfirms the latter). Strongly underdetermined alternative theories are 'empirically irresolvable'. In this essay I mainly analyze the existence and implications of *observational* indistinguishable – or irresolvable –conflicts (and will frequently drop the adjective 'observational' if no confusion is threatening). Nevertheless, in certain classes of cases of 'irresolvable conflict of assertion' one does not expect physical facts and experience to settle the matter. They appear thus 'irresolvable' or indistinguishable in that the reasons advanced for each of the conflicting alternatives are of equal standing, as is the case, for instance, with the four Kantian *antinomies of pure reason*. A more recent example perhaps are competing set-theoretical accounts of arithmetic. I shall call such conflicts of assertion *rationally* irresolvable.

1.2 Significance of the Thesis

Suppose the phenomenon of underdetermination or 'empirical irresolubility' is genuine and sufficiently widespread in the sciences.

Which epistemological and ontological consequences may be legitimately drawn? Most widely discussed is the impact of the underdetermination thesis on the issue of *scientific realism*: how can data warrant belief in what a particular theory says about the unobservable, if one knows there are rival theoretical accounts, which have identical empirical content? Before I turn to this particular matter, I like to point out a number of less frequently mentioned consequences.

First of all, the construction of conflicts of assertion is a genuine philosophical *Denkfigur* or method. The successful construction of an alternative that is not excluded by the grounds advanced for (or is compatible with) a *prima facie* well-established thesis or assertion, challenges the justification of the thesis. For instance, the skeptical alternative that an individual may be actually dreaming, while having apparently waking experiences, has frequently been used to target the rather plausible thesis that sense experience is a source of knowledge of the world. Irresolvable conflicts ('empirical' or not) were frequently taken as a *prima facie* strong inducement to *scepticism*. We cannot claim knowledge that p, or at least that our belief that p is rational, if $\neg p$ appears as justified as p. The use of the *Denkfigur* is not, however, limited to the sceptical tradition. Richard M. Hare argued against the *existence* (or rather the *intelligibility*) of objective moral values on the ground that two worlds like ours, which only differ over the inclusion of objective values, all other things being equal, are observationally indistinguishable (see section 1.5). Quine argued against the existence of propositions on the ground that empirical irresolubility with respect to alternative translation schemes for a foreign language demonstrates fundamental problems for the individuation of propositions (see chapter 6). As a philosophical method, 'irresolubility'-based arguments are *topic-specific*, local and do not rely on general existence claim about alternatives.

One potential casualty of empirically irresolvable conflicts in the sciences is a popular and plausible analysis of (physical) *lawhood*. A law, according to F. P. Ramsey and D. K. Lewis, is either a high-level postulate or a consequence thereof in the "best systematization" of the ideally complete account of our experiential knowledge. A hypothetical multitude of alternative accounts, all equivalent in saving the phenomena, appears to make lawhood subjective – provided there is not by some necessity a large common core between rival theoretical accounts. Hence one is led to doubt Lewis and Ramsey's analysis of lawhood.

Another potential consequence of observational indistinguishability is the impossibility to naturalize truth, that is to explain (scientifically) truth as 'correspondence' through the notion of linguistic reference by appeal to facts, causation and natural laws. Truth is usually explicated through the notion of satisfaction, and hence relies on a notion of reference Field (1974). A reference 'scheme' is a systematic assignment of denotations to names and predicates. Since many divergent reference schemes make the sentences of the theory come out true, the theory fails to determine the reference of its terms. H. Putnam offered a well-known argument in support. Our comprehensive theory, insofar as it can be formulated in first-order logic and is consistent in syntactic terms, has more than one model. Given one model all its isomorphic images satisfy the same theory (see chapter 4). Hence even though a theory may be comprehensive and epistemically ideal and satisfy all physicalistic requirements one may impose on the terms in the theory's vocabulary it cannot explain or determine the notion of reference for its terms, Putnam claimed. Note, empirical irresolubility or strong underdetermination is not the only premise in deriving those far-reaching consequences, nor is it perhaps the most controversial.

An important motivation for examining the thesis of strong underdetermination of theories is the potentially negative impact on the very possibility of scientific methodology. It touches on the essence of science, since science's requirement of testability and its distinct methodology marks it sharply off from other intellectual enterprises. (The difficulty to be outlined here is independent of the viability of scientific realism – at least to a first approximation.[5]) D. Mayo worries that on account of (methodological) underdetermination "criteria of success based on methodology and evidence alone undermine choice

---

[5]In the context of an inquiry into the conditions of the reliability of inductive methods Earman stresses that the present issue "transcend[s] questions about the status of theoretical entities" (Earman, 1993, p. 20).

[between alternative hypotheses]", rendering *any* attempt to "erect a methodology of appraisal" of theories hopeless (Mayo, 1996, pp. 174, 176). Without a methodology of appraisal, it seems, scientific change is subject to other forces than pure rationality and not just in peripheral ways. Anything goes in the sciences? Laudan, after examining various versions of the thesis, concludes "that no one has yet shown that established forms of underdetermination do anything to undermine scientific methodology as a venture, in either its normative or its descriptive aspects" (Laudan, 1996, p. 53). To bring out the difficulty more sharply, *eliminative induction* may serve as a model. The scientist starts out with a set of explicitly given, 'seriously' entertained alternative accounts for a set of observations; devises experiments and tests with the aim to weed out the false ones among them, and if lucky finds one theory among the original set that withstands these and further tests. One tends to think of the scientist as justified in either believing the surviving theory (if there is one), or at least in "accepting" it, since his belief is generated by applying the proper method of inquiry. When the surviving theory eventually fails, the process restarts with a new initial set of strengthened theories. Discounting the familiar problems raised by the weak underdetermination of a hypotheses, strong underdetermination blocks the method in two ways.

First, if there are fully observationally equivalent alternatives in the original set, then if one of them survives all survive (I ignore the dependence of a theory's predictions on auxiliary hypotheses for the present purpose). Since the empirically equivalent alternatives may include incompatible theoretically statements the method cannot issue in a belief-verdict, not for one of them nor for the whole bunch. The situation is less dramatic when inductive elimination is taken to produce, in favourable circumstances, merely an accept-verdict. Arguably, a scientist can accept a set of theories (as instruments, as empirically adequate) even if some among them are incompatible on the theoretical level. (However, acceptance seems a more temporary, transient state of mind than belief when it comes to scientific practice.) Second, suppose none of the theories in the initial set of "seriously entertained alternative accounts" has an empirically equivalent partner in the set.

If the strong underdetermination thesis is true, then each has nevertheless empirically equivalent partners outside the set. The method would justify belief, under favourable circumstances, in the unique surviving theory. Yet, the justification is achieved only by excluding certain theories from the initial set. Had an empirically equivalent partner of the surviving theory been included, instead of the initial limited choice, the method would not have justified belief in it. The justification therefore is phoney, except if there are good inductive reasons for thus restricting the original set. If no empirically equivalent alternative to the surviving theory is explicitly known, the scientist surely has a right to proceed without waiting for it to be discovered and included. This is a rare case, though, since sometimes genuine alternatives are known, and there are known "algorithms" for constructing empirically equivalent, ontologically non-equivalent alternatives. Should one exclude these as 'frivolous' theories and "cheap tricks", as has been suggested? I will return to this question, but it seems safe to say for the moment that if someone were under the illusion that scientific methodology is captured by our simplistic version of inductive elimination and its upshot is belief, then underdetermination puts an end to it.

Scientific realism, it is frequently argued, is incompatible with the thesis of strong underdetermination, and if the latter, properly formulated, is true, the former must be rejected; hence the philosophical significance of the thesis. Indeed, something is amiss with belief in a given class of objects x if (i) an (perhaps equally simple) alternative is available that saves the phenomena without x or with y and z instead of x, (ii) belief in which is as justified as the belief in the original class. The focus of the present discussion is on the general shape of antirealist arguments based on empirical irresolubility between theories. Boyd holds that strong underdetermination or 'empirical irresolubility' is a key premise in a "simple, and very powerful argument that represents the basis for the rejection of scientific realism by philosophers in the empiricist tradition" (Boyd, 1984, p. 42). Paul Horwich argued that the general (strong) underdetermination of theory by evidence in the sciences is the best argument for instrumentalism with respect to the entities and states postulated: theoretical claims are not proper objects of belief (observation statements are), they do not "represent" anything. Needless to say, Horwich rejects the premise of the argument (Horwich, 1991, p. 11).

Richard Giere holds that the (alleged) phenomenon of empirical underdetermination in the sciences strongly push the naive realist to adopt a standpoint-relative or 'perspectival' version of realism (Giere, 1993, p. 21). Since the data plus (ideal) scientific methodology cannot 'determine' the theory, our interests and predilections, hence some element of *cultural bias*, take up the slack.[6] Underdetermination makes this truth explicit. Giere too, however, has reservations about the validity of the major premise, which drives the argument. Crispin Wright goes a step further and concludes that the realist should adopt an epistemic notion of scientific truth or else is driven into the "absurdity" of admitting the possibility of an empirically adequate, simple theory that may nevertheless be false in the correspondence sense of truth (Wright, 1992, p. 25). He is working from a notion of realism that goes back to Michael Dummett, who characterized realism as "the belief that statements of the disputed class [sentences in the theoretical vocabulary – TB] possess an objective truth-value, independently of our means of knowing: they are true or false in virtue of reality existing independently of us" (Dummett, 1978, p. 146).

Michael Devitt, in a recent, widely read defense of realism, argues that the difficulty for the doctrine of realism does *not* arise by an inductive inference from numerous and suggestive examples of underdetermined theories in the sciences. Rather it is properly located at the philosophical meta-level in the context of modern skepticism:

> The skeptical strategy is to insist that the Realist is justified in his belief only if he can give good reasons for eliminating alternative hypotheses to Realism. Yet, the skeptic argues, the Realist is unable to give such reasons, as he (the Realist) should be able to see. Realism collapses from

---

[6]Various theses of underdetermination of theory by data play an important role in social approaches to knowledge and justification; see, for instance, Longino (2002, pp. 124f), especially the discussion of Ph. Kitcher's 'research pragmatic' objections against the significance of underdetermination (pp. 62f). The issue is peripheral to my concern.

within. [...] The Cartesian doubt reflects the truth of 'underdetermination'. The evidence of our senses undermines our views of the world: other opinions are compatible with the evidence. This truth is supported by our best modern science. (Devitt, 1997, p. 62)

There is then, if we follow Devitt's way of setting up the problem, no objective justification for selecting one from a variety of conflicting metaphysical interpretations of the totality of our experiences as the true or most likely one. These conflicting interpretations include Berkeleyian immaterialism, solipsism, Descartes' *genie malin*, the machinations of vat-operators or endless variations on similar science fiction themes. The common sense postulation of independent causes operating in space-time is just one interpretation among incompatible other ones of what we actually observe or potentially could observe. Justification in Devitt's analysis, incidentally, takes the form of induction by elimination (of all but a realist conception of the external world) based on "evidence of our senses". He treats "views of the world" methodologically in analogy to empirical theories.

Of the many proposed formulations for realism he chooses the following statement:

> Tokens of most current observable common-sense and scientific observable physical types objectively exist independently of the mental. (Devitt, 1997, p. 24)

The claim asserts something about the existence of cats, say, not about this or that specimen, hence the type-token distinction. By introducing the independence claim Devitt wants to avoid any reference to truth by correlation or correspondence in the realist thesis. Although the general intent is clear enough, this particular formulation of the doctrine suffers from an ambivalence in the concepts "the mental" and "independence". For committed naturalists, and Devitt is among them, the statement is trivially true: the mental refers to neurological processes fairly well localized inside the human skull. (At best, it rules out telepathy and the like.) If however "the mental" is understood in a general, non-committed sense, that includes, for example, idealist conceptions of *Bewusstsein*, than the statement is equivalent to the blanket metaphysical denial of all such alternative conceptions. In either case the doctrine seizes to be a hypothesis in the usual (and intended) sense. The basic realist intuitions are better served by a formulation that centers around the justificational independence of the entities in question from particular historical, sociological, and linguistic contingencies of the of the discovery and acceptance of the "embedding theories" (or conceptual systems) that define principles and laws which govern those entities into account:

Real Tokens of most current observable common-sense and scientific observable physical types objectively exist independently of the way and context in which those types (or their embedding theories) have been framed or discovered.

Realism can be formulated with regard to many kinds of systems of entities: the external world, the past, propositions, moral values, intensional states, numbers, and in each case empirical irresolubilty arise in much the way. Realism about *sentence meanings* would be undermined by the indeterminacy of translation, an argument I will examine later (chapter 6). Existence claims about *past events* appear to be strongly underdetermined by Russell's "hypothesis". One has arguably *direct access* of a mostly non-inferential basis one's own past experiences, at least for the very recent past, under normal conditions. However, the memory traces are few, incomplete and inexact, and they extend only so far. Russell remarked that it is not inconsistent with our observations and subjective memories to maintain that the world sprang into existence five minutes ago, complete with all traces of the past, including memory traces Russell (1921). Underdetermination casts its shadow even over the common sense view of the passage of time. The logical possibility that time has the topology of a *circle*, instead of that of a line, with history caught in perpetual repetition, is discussed in Horwich (1982, pp. 74–76) and Newton-Smith (1978). Newton-Smith affirms the theoretical non-equivalence of the two stories. Paul Horwich maintains that these alternatives are "potential notational" variants and that the conventional adoption in practice of one view entails *a priori* the rejection of the other (see section 5.3). S. Shoemaker investigated the possibility that all motion in the material world "freezes" for a finite duration and then proceeds where it left. Since the observer's perceptual and cognitive states freeze as well and no traces of the universal freeze remain, we assume, one is led to think that this scenario is observationally indistinguishable from the common sense view of the "gap-free" flow of time Shoemaker (1969). Thus there may be no good reason to choose the one over the other.[7]

How exactly does the conflict between scientific realism, in the form of the thesis Real, and underdetermination arise? The claim that a thesis of strong underdetermination presents a significant obstacle for the defense of scientific realism depends not only on it being true but more than is perhaps usually appreciated on the thesis' fine print and its scope. For instance, suppose only scientific theories of a particular mathematical type are prone to have fully empirically equivalent rivals, and that few theories actually employed are of this type. The restricted thesis of underdetermination would hardly be incompatible with the realist interpretation of other types of theories. (There is little to support this fantasy; theories as structurally different as quantum mechanics and general relativity both appear to have fully empirical equivalents.) Or, it may turn out that we have reason to believe that any theory (of sufficient "critical mass") has at best only a very limited numbers of observationally equivalent alternatives. Again, a restricted thesis of underdetermination like this one would hardly be incompatible with scientific realism. Since the number of alternatives is strictly limited they would likely have a lot of theoretical structure in common. On matters on which the alternatives contradict each other, a scientist may be wise to withhold belief and remain *agnostic*.

Actually, requiring an agnostic attitude from scientists in the face of underdetermined alternative theories is untrue to scientific practise and perhaps unwarranted. The scientific community when faced, for instance, with a novel, successful theory and a few of its observation-

---

[7]E. Sober ranks both views equally in point of statistical simplicity, but the non-orthodox view postulates gratuitous entities, extra hidden years, and hence has a less "parsimonious ontology" (Sober, 1996, pp. 178–179).

ally underdetermined alternatives does not, as a historical matter of fact, typically opt for either skepticism or general retreat from realism in interpreting theoretical claims. Rather, the community 'splits' temporarily into a few factions or research programs, each endorsing one genuine rival as descriptively true and rejecting the others. A genuine rival theory to the novel (primary) theory is an alternative that *compensates* for any of the former non-standard, unusual theoretical aspects by entailing a hypothesis that members of the scientific community consider to be true. This kind of theoretical compensation typically avoids or solves a (perceived) conceptual problem of the novel theory. Or, the alternative has a theoretical feature, which some scientists think essential for 'understanding' and 'explanation' provided by the theory. Inconsistency with certain highly confirmed background principles can become a (evidential?) reason for rejecting an alternative in favor of an empirically indistinguishable rival. A *genuine* alternative offers a conceptual advantage (not to be confused with "progressiveness" in Laudan's terminology) as measured by its conservativeness over the novel, primary theory. The remarkable thing is that although the commitment of each faction is *dogmatic* (in the modest sense of being non-evidential) the response of the scientific community to the underdetermination of a successful theory as a whole is rational. Each group proceeds 'dogmatically', but the community as a whole does not. It is advantageous for a *risk-averse* scientific community in the search for 'truth' to pursue different strategies in the face of theoretical uncertainty generated by the underdetermination of theories. The *tempered pluralism* with regard to belief, advocated here, presupposes a naturalistic view of the scientific enterprise, modelling the scientific community as a set of rational agents who adjust their views in the face of uncertainty (taking a leaf from Ph. Kitcher).

Scientific realism has been presented mainly as a thesis about truth (bivalence) or reference (compare statement Real). But realism has an *epistemological component* as well, it includes a claim about rational belief or the possibility to acquire justified beliefs about the unobservable by a well-defined process (a method). If a statement about unobservables is true, one should be able to find specific evidence for the statement; the piece of evidence should properly 'track the truth'.

It becomes plain then why many perceive a conflict between scientific realism and strong underdetermination. A strong underdetermination thesis seems to be directly in conflict with the epistemological component of realism. If true, the thesis would deprive the inquirer of the possibility to have *observationally* justified beliefs about unobservable events, objects and processes.[8]

In order to pinpoint the conflict some stage-setting and simplification is called for. Suppose 'conf(e, T)' stands for (theory) 'T is supported by a set of observation sentences e', where conf() is a *qualitative*, absolute, all-or-nothing concept of support. The class of all (non-analytically) true observation sentences entailed by T is the *truth content* of T, $c_t(T)$; similarly wrong predictions form T's falsity content.[9] For T to be supported requires minimally that its falsity content be empty and its truth content be non-empty. The *epistemic* component of scientific realism may be expressed in the present context as an inference from support for T to justification of belief; i.e. if T is supported by e than one is justified to believe T. Under the assumption that a strong underdetermination thesis holds (premise 3 below), the *prima facie* difficulty for realism's epistemic commitment now takes this form:

1. T_0 is a theory such that $c_t(T_0) \neq \emptyset$ and $c_f(T_0) = \emptyset$.

2. For every theory T: $(c_t(T) \neq \emptyset$ and $c_f(T) = \emptyset) \rightarrow$conf($c_t(T)$, T).

3. For every theory T there is another theory T' such that $\neg(T \& T')$ and $c_t(T) = c_t(T')$ and $c_f(T) = c_f(T')$.

4. Hence, conf($c_t(T_0)$, T_0) *and* conf($c_t(T_0)$, T').

5. From conf($c_t(T_0)$, T_0) infer 'belief in T_0 is justified'.

6. From conf($c_t(T_0)$, T') infer 'belief in T' is justified'.

---

[8]The force of strong underdetermination is essentially independent from the species of realism, 'entity realism' or 'structural realism' proposed by J. Worrall for instance, one adopts.

[9]For details see p. 92; the concept of truth content is due to K. Popper.

7. Therefore, belief in T_0 is justified *and* belief in T' is justified.

But one cannot rationally believe two contradicting claims to be true simultaneously, given the assumptions about the all-or-nothing character of conf(). There are obviously three direct ways to avoid the conclusion: reject the account of support, a naive qualitative account of confirmation in favour a sophisticated methodology (premises 1 and 2); reject (the special form of) the underdetermination thesis expressed in premise 3; or reject the epistemic component of scientific realism in the form of premises 5, 6. All three responses can be traced in the literature on the subject.

The reconstruction does not yet fully bring the conflict between a strong underdetermination thesis and the epistemic commitment of scientific realism to the fore. To support anti-realism, a thesis of strong underdetermination should say that *anything* a scientist wants to claim about unobservable processes and entities can be contradicted without impugning the capacity to account for the data. 'Accounting for the data' is too weak an expression. Rather, if the theoretical statement $\neg S$ were true instead of S, the evidence would still fully justify belief in S. What turns a datum, an observation, into evidence is methodology or a theory of confirmation. Suppose M is a rationally defendable, ideally complete methodology (which may vary from one research area to another; I do not take this dependence into account). For the epistemic component of the realist doctrine to be false something like the "universality" thesis below has to be true:

U   For every purely theoretical statement S of any theory, or system of belief, of sufficient 'critical mass' and scope, regardless of how well empirically supported (relative to M), there exists an alternative theory, or system of belief, which entails $\neg S$ and is as well empirically supported (relative to M).

If U were true, the evidence remains undisturbed in a world in which $\neg S$ is actually true instead of S.[10] The *ideal* (inductive) method M

---

[10] I refer to the statement as 'universality thesis' to emphasise the universal quantifiers and its necessary generality; it is an expression of strong underdetermination.

does not track the truth. A couple of remarks are in order. The qualifier 'of sufficient critical mass and scope' reflects an interest in scientific theories (not single theoretical claims), and a Quinean concern with underdetermination of the total system of belief. Note an ambiguity in that the notion of evidence refers to the embedding theory of S and $\neg S$ respectively, not to S in isolation; this is realistic but the formulation can improved. It is useful to draw a distinction as to whether U is formulated with respect to equality of deductive observational content, or with respect to equality of relevant evidence.

The phrase 'there exists' a theory in the statement of U is surely problematic. If it means that there is a *logical possibility* that an alternative with such and such theoretical features exists, although none is known as a matter of fact, the plausibility and strength of U quickly dwindles. There are too many worlds, which we cannot rationally exclude as logically (conceptually) impossible. Many find this reading of U, with its gesture towards Cartesian skepticism, uninteresting.[11] However that may be, the phrase 'there exists a theory' can well be taken in a constructive way: it is not too difficult to construct incompatible alternatives to any given theory. I take up this thorny issue in chapter 4.

It is obviously far from clear that the strong thesis U is true or even plausible. Thus when Putnam writes matter-of-factly of the "phenomenon" of underdetermination he understates the claim in question and underestimates the strength required for U. If U is false, *although* a more restricted variant perhaps is true instead, Horwich's (and other's) claim that strong underdetermination on the level of scientific theories would wipe out scientific realism is premature.

Many scientific realists feel the bite of strong underdetermination in a different place. They view scientific realism as an *empirical* hypothesis, which is confirmed or explained by the success of (modern) science. The enormous growth in the capability for prediction and control cannot plausibly be explained, it is argued, as being right by pure

---

[11]Kitcher helpfully suggests that attempts to interpolate underdetermination claims to theories beyond the borders of physics merely "raise skeptical possibilities for which philosophers are notorious" (Kitcher, 2001, p. 195).

chance, and hence the success is very likely due to the approximate truth of our theories. An underdetermination thesis, however, if true, even in a less restricted form than U, seems to show that success in prediction and control of natural processes is not the unique pointer to the truth of our claims about unobservables. Empirical success cannot well be explained by the truth-likeness of our currently accepted, mature theories, since there are equally successful rival, but presumably *false* theoretical structures. In this case, the significance of strong underdetermination lies in its power to undermine a particular defense of scientific realism, the 'no miracle argument' or the 'inference to the best explanation'. The potential defeat of a particular argument for scientific realism is of course a far cry from undermining our confidence in realism itself.[12]

1.3 Quine, Realism, and Underdetermination

Willard V. Quine had forcefully argued that the existence of irresolvable conflicts has a profound impact on metaphysics, epistemology and semantics. Quine held that the facts, however rich and varied, together with logic and scientific methodology cannot settle our picture of the material world in important respects.

> What the empirical under-determination of global science shows us is that there are various defensible ways of conceiving the world. (Quine, 1992a, p. 102)

Similar statements are scattered throughout Quine's work and I will revisit them in later chapters. Subtracting certain trivial ways in which the world can be alternatively described we are left with an astonishing although perhaps not novel claim. 'Empirical irresolubility' of two systems of global science seems to help drawing a line between what

---

[12]For other difficulties related to the 'no miracle argument' see the critical examination in Howson (2000, pp. 35–60). Underdetermination is always a threat if the doctrine of scientific realism is understood as a (high-level) empirical or testable claim about the nature of our theories.

is part of the conceptual scaffold and what is due directly to nature, between that which is "variable" and human and what is tightly bound to observed regularities and physical fact.[13] In the previous section I have examined the apparent tension between realism and what I have called empirically irresolvable conflict or strong underdetermination. Quine is one thinker who *denies* that there is a significant tension. Quine subscribed to a realist view of physical – and abstract – objects (Quine, 1966, pp. 228–230). He is also the thinker who harnessed empirical underdetermination in many ways to derive much discussed claims regarding ontology and the non-intelligibility of traditional sentence meaning. There are, Quine argued, conflicting alternatives to whatever ontology is inherent in a system of belief or theory a speaker holds. The resulting conflict is empirically irresoluble, in the sense that alternative ontologies can be attributed to the speaker and still preserve all accepted objective, factual truths.

How does the commitment to general empirical underdetermination and irresolvable conflict of assertion, and the realistic outlook go together? If Quine did not locate the significance of the existence of empirically irresolvable conflicts in a threat that it poses to realism, then what is its significance? Before I turn to these questions it is appropriate to review Quine's reasons for the empirical underdetermination of most systems of belief. Below is one characteristic statement, representative for a great number of scattered remarks spanning four decades. The expression "surface irritations" refers to the external visual, tactile, etc. stimuli.

> Actually the truths that can be said even in common-sense terms about ordinary things are themselves, in turn, far in excess of any available data. The incompleteness of determination of molecular behavior by the behavior of ordinary things is hence only incidental to this more basic indeterminacy: *both* sorts of events are less than determined by our surface irritations. This remains true even if we include all

---

[13]Davidson (1974) has questioned the intelligibility of a scheme–content distinction. For a criticism of his views on this point that I have found persuasive, see David Papineau's *Reality and Representation*, Oxford: Blackwell, 1987.

past, present, and future irritations of all the far-flung surfaces of mankind, and probably even if we throw in an in fact unachieved ideal organon of scientific method besides. [...]

Peirce was tempted to define truth outright in terms of scientific method, as the ideal theory which is approached as a limit when the (supposed) canons of scientific method are used unceasingly on continuing experience. But there is a lot wrong with Peirce's notion [...] For, as urged two pages back, we have no reason to suppose that man's surface irritations even unto eternity admit of any one systematization that is scientifically better or simpler than all possible others. It seems likelier, if only on account of symmetries or dualities, that countless alternatives theories would be tied for first place. (Quine 1960b, pp. 22, 23; emphasis by Quine)

Quine advances here two reasons for a general "indeterminacy of determination" of theory by observations: (i) an epistemological argument from the nature of scientific methodology; and (ii) a semantic argument from the residual existence of "symmetries or dualities" in an ideal, best systematization of all observations. The latter argument is supposed to support the thesis independently of (i), i.e. under the assumption that scientific method is potent enough to eliminate all but out one theory in the long run.[14] The reference to symmetries is to basic symmetries in the laws or objects postulated by the ideal theory: bilateral, permutational, rotational or translational symmetries. Alternatives that differ in the assignments of values to magnitudes which are subject to such symmetries cannot be discerned by observation. Duality is a well-known property of sentential logic: the *dual* A' of a statement form A which involves only negation, 'and', and 'or' (but no 'implication' or 'equivalence' signs) arises from A by systematic replacement of 'and' by 'or', 'or' by 'and' and every statement letter by its negation. Then A' is logically equivalent to $\neg A$. Similar dualities hold

---

[14]The semantic argument is not present in a similar remark on the underdetermination of theories in "Two Dogmas of Empiricism".

for theorems in first-order predicate logic, the algebra of sets and plane geometry. The suggestion of that brief remark then is that cyclic permutations of this kind in the *theoretical* statements of one "best" ideal systematization of all our knowledge result in genuine alternatives that are empirically equivalent, assuming that the language of science has been properly reconstructed. Differences due to cyclic permutations are (it is said) beyond the reach of scientific methodology, since they do not affect the requirements of simplicity and no observations can make a difference. Generally speaking, systematic permutations within the universe of discourse, along with 'compensatory' reinterpretation of the basic predicates, preserve all true observation sentences that can be deduced from the theory. The general thesis, of course, goes under the name of *ontological relativity*, namely relativity of the implicit ontological import of a speaker's theory of the world to the way this theory is translated into another, the theory of the hearer or linguist.

The *methodological* argument, (i), is the one Quine relied on most frequently and is perhaps the more plausible. Quine's notion of methodology (for epistemological purposes) is very close to classical hypothetico-deductivism. The theory of testing and confirmation, which hypothetico-deductivism has as its base, leads to *confirmational indeterminacy* (see section 3.2). Extra-evidential criteria are needed to rule 'theory choice', like the requirements of falsifiability ("refutability"), non-ad-hocness ("generality"), the degree of precision of predictions, theoretical economy ("modesty") and, of course, simplicity. These criteria have proved themselves by their utility in past scientific practice, are fallible and apt to change. In addition, and unlike most other methodologists,[15] Quine emphasized the requirement of "conservativeness" in theory change: the best among rival new theories is the one that preserves the most of the theoretical (and empirical) claims of its forerunner. The main reason behind "conservativeness" as a requirement is the need to bring Quine's doctrine of the empirical *revisability of logic* in line with the overall historical stability of logic and mathematics despite profound changes in the course of the

---

[15]R. Boyd's account of inter-theoretical relevance of evidence turns on this requirement for its overall argument, see section 3.5.

modern sciences. Of these six methodological requirements only two are relevant for the kind of synchronous theory choice problem that a potential multitude of "best" global theories presents us with: simplicity and conceptual economy. Both criteria are imprecise and lack substance. It is therefore doubtful that these two criteria will eliminate all but one from a potential array of ideal systematizations of all our knowledge. Quine concludes from the methodological argument that an "indeterminacy of determination" by observation at the level of total science becomes likely.

Quine briefly considered the skeptical turn and the retreat into a kind of instrumentalism inherent in strong global underdetermination:

> It sets one wondering about truth. [...] Can we say that one [of alternative systems of the world – T.B.], perhaps, is true, and the other therefore false, but that it is impossible in principle to know which? Or, taking a more positivistic line, should we say that truth reaches only to the observation conditionals at most [...]? (Quine 1975, p. 327)

Over the years Quine tackled these questions from various angles. One could, should the rare case arise, switch pragmatically between two or more empirically equivalent systems. That is, one adopts one or the other as the circumstances may require, like physicists switching between a description of micro-phenomena in terms of a wave or a particle picture. Another proposal to avoid irresolvable conflict between assertions, due to observational equivalence, is to rename terms in the (class of) statements which contradict each other (see section 5.1). Thus Quine hoped to avoid the loss of objectivity in our description of nature and justified beliefs. Moreover, it is an open question, according to Quine, if "genuine" underdetermined alternatives are *bound to exist* beyond a potential accidental failure to make them commensurate to our system. That a theory is bound to have alternatives can mean different things. Logical necessity? A *proof* of the existence of alternatives, or the construction of an algorithm to generate underdetermined alternatives to a given theory? Necessity can refer to the capability of human minds to create a stream of variant theories, or its *incapability* to find ways to reconcile two given alternatives. The claim to necessity may be derived from methodology: no unique representation of natural processes can be inferred. If we follow Quine, observational indistinguishability with respect to our "total science" exhibits "what is humanely possible" in scientific inquiry Quine (1975).

Why did Quine not appreciate strong underdetermination as a threat to realism? First a word about his brand of realism, which is best characterized as a version of W. Sellar's "hypothetico-deductive realism". Hypothetico-deductive realism emerged as an alternative to empiricist (sense-data based) accounts of statements referring to external objects. It erases metaphysical and epistemological boundaries between naive realism and scientific realism. The justification of belief in *all* kinds of objects is taken to follow the hypothetico-deductive model. The 'material-object-hypothesis' is thought to be justified basically in the same way the 'proton-hypothesis' is, either by way of minimal inductivism or as the "best possible (deductive) explanation" of our observations. Hypothetico-deductive realism is a superior analysis of common sense existence claims when compared to (i) the classical empiricist account, which is entangled in problems of reduction and meaning; (ii) to naive realism, which creates compatibility problems with postulated objects of the sciences by taking certain observational predicates of objects as logically primitive; and (iii) to "metaphysical" realism, which raises the paradoxes of the knowability of things in themselves.

Quine's answer to why there is no tension between strong underdetermination on the level of global science and realism is, in one word, his commitment to *naturalism*. Philosophers, like scientists, are engaged in reworking what is currently accepted in the sciences and by common sense, famously "there is no first philosophy". Among the possible and actual multitude of systems, we are partial to the one we are actually working in.

> In my naturalistic stance I see the question of truth as one to be settled within science, there being no higher tribunal. This makes me a scientific realist. (Barrett and Gibson, 1990, p. 229)

1.3. Quine, Realism, and Underdetermination

For these reasons Quine felt that he can accept both a fully fledged realism and the existence of empirically irresolvable conflict. It may seem as if Quine should have arrived at a *structuralistic view* of our global theory, like some earlier empiricists. What is objective in our theories of the world, common sense or scientific, are structural relationships. Yet, Quine shies away from taking this course, since we cannot help but interpret the system of entities we are actually presupposing in researching the world, i.e. the posits of the sciences, realistically. Not only is there no first philosophy, there is no 'cosmic exile' either. His is an epistemological structuralism, not an ontological one. Roughly put, Quines seeks to escape the difficulty for the epistemic component of realism created above, by rejecting Real and by moving from theories proper to 'natural science' and applying a different standard of support for the latter.[16]

For all its virtues hypothetico-deductive realism is mired in difficulties. First, it is *intrinsically* liable to underdetermination arguments since it is based on hypothetico-deductive methodology, which engenders confirmational holism. Of course, not everyone will see this as a fatal shortcoming of the method, and Quine is among them. Another worry concerns the validity of the analogy with hypothetico-deductivism in the sciences. In the sciences "posits", along with the laws governing them, are tested against observation sentence that are independent of the hypothesis under test. It is different with the 'body-hypothesis' or "thing-theory". In contrast to the posits of micro-theories, "thing-theory" has penetrated down to the level of perceptions: I perceive a desk, not a rectangular shaped, brown patch. The latter impression usually has to be isolated afterwards by shifting the focus of attention. We describe phenomena spontaneously using a language of particulars located in space and time. Empirical generalizations of the everyday kind thus seem to presuppose a common sense or naive realism. Therefore the analogy to scientific theories is methodologically misleading. There may be well be nothing to *test* the common sense system of bodies, the "thing-theory" against. The

---

[16]See Quine (1992b) for his discussion of structuralism, section 6.5, and Bonk (2004) for a more detailed account.

material-object-hypothesis cannot be refuted or falsified by observation, at least not in the customary sense of falsification. Physical bodies and generalizations based upon them already pervade observation, and there is no alternative idiom. Finally, Quine assumed that there is such a thing as an (evolving) homogeneous, unique scientific "system of the world", which after reconstruction in first-order logic and set-theory indicates our ontological commitments. This is an idealization, to be sure, made to make an epistemological point about the genesis of this edifice out of the 'torrent of stimuli'. It needs to be stressed, however, that the assumption is a *gross* idealization: at any one time in the recent history of the sciences a multitude of theories, models and programs co-exist which are mutually inconsistent. Nor is this multitude mutually (ontologically or theoretically) reducible in any uncontroversial sense Moulines (1997). If the practice of the sciences shows a diversity like this there is little point in pretending there is a unique, global 'system of the world'.

1.4 No Quick Solutions

Among the objections to the thesis and the significance of strong underdetermination, there are a few which I like to dispel at the outset, since I believe they misrepresent the issue in one way or another. One objection says that there are not enough examples of irresolvable conflicts of assertions, suggesting that the problem is speculative and 'scholastic'. A second objection points out that although there are good examples of empirical underdetermination, they all pertain to theories of physics and there are none from other sciences, suggesting that the problem has little significance beyond the methodology of physics. Yet another objection has it that the best explanation for a lack of success in discovering an observationally equivalent alternative to an orthodox theory is that such alternative theory simply does not exist. A final objection to be considered here, or rather re-formulation of the issue, emphasises that underdetermination is an instance, or extension, of the familiar problem of induction. Depending on the thinker, the latter has either been solved or is dismissed as hopeless. Unsurprisingly perhaps, these objections tend to be emphazised by (scientific) realists.

The metaphysical and epistemological significance of empirically irresolvable conflicts has been dismissed on the ground that there is not enough reason to start worrying. For instance, Crispin Wright states: "However there is, so far as I know, *absolutely no compelling case* for believing in Maximal Underdetermination as formulated"(Wright, 1992, p. 25; emphasis by C. W.). What makes a compelling case? If proof is required, than indeed there is no such case for believing in a *systematic* failure of our theoretical beliefs to be uniquely determined by the (totality of) relevant facts. Yet, proof is a requirement that is rarely satisfied in philosophy. On the other hand, there are both several noteworthy examples from science (see chapter 2, compare Earman, 1993), and general arguments to lend *prima facie* strong support to the claim in one form or another. While these cases are not meant to be the basis of a sweeping generalization, as has occasionally been imputed, they should suffice to motivate reasonable doubt about the observational 'uniqueness' of some of our theories. It is misleading to suggest that the *only* argument for the existence of empirically irresolvable conflicts is an inductive generalization from cases.[17] The objection appears to neglect, for instance, the systematic arguments Quine advanced in a number of papers, which have not been defeated. Reichenbach's program of analyzing theories and languages by isolating meaning postulates or coordinative definitions is a potentially *unlimited source* of empirically irresolvable conflicts of assertion. The fact that Reichenbachian conventionalism has been discredited is not relevant here, since conventionalism was proposed as a response to the problem of empirically irresolvable conflict. Even if it fails as an analysis of the problem, the examples if not the mechanisms of generating alternatives that Reichenbach was fond of exhibiting are still valid. Thus M. Friedman comments: "What happens if we reject the conventionalist strategy and maintain that the alternative, empirically equivalent theories in question should be taken at face value [...]? To

---

[17]H. Putnam, for instance, appears to argue mainly from suggestive examples, see Putnam (1978, p. 133), so does van Fraassen (1980a).

avoid skepticism we must hold that [...] there are methodological principles or criteria [...] that are capable of singling out but one theory from a class of empirically equivalent theories (or at least capable of narrowing the choice considerably)" (Friedman, 1983, p. 267). Those who see no compelling case for the underdetermination of theories, must have such 'methodological principles or criteria' in mind; I review some salient proposals in chapter 3.

The appeal to cases by proponents of underdetermination is sometimes countered by the remark that although *there are* cases of underdetermination, there are no "good" or philosophically "interesting" examples. Alternatives constructed from non-standard logical systems, or from algorithms hoisted on an otherwise perfectly parsimonious theory etc., are dismissed wholesale as "logico-semantical" trickery (L. Laudan). I am sympathetic. Yet, on the other end of the spectrum of cases are both current quantum theory and general relativity theory, which appear to be empirically underdetermined with well-known alternatives. These are fundamental theories of universal scope and great predictive power. In any case, the qualification "good" or "interesting" points to an explicit *criterion* that eliminates all but one empirically equivalent alternative, and a justification for why the theory selected according to the criterion is more likely to be true. The critic thus in fact accepts observational underdetermination of theories, but believes he has a ready solution at hand. Unfortunately, where criteria are made fully explicit they are rarely sufficiently motivated as a rational basis for the dismissal of an alternative Kukla (1998).

A variant of this objection stresses that there are no examples of observationally irresolvable 'conflicts of assertion' in biology, geology, chemistry, etc. The 'best' examples come solely from *one* branch of natural science: physics. Since hypotheses in other areas of science are not inflicted by underdetermination a generalization from one or two physical theories hardly seems justified (for instance, Miller, 1987, pp. 424f; Kitcher, 2001, pp. 195f). If valid, this objection would tend to shrink the significance of empirical irresolubility to a point where it becomes a mere symptom of *lacunae* in our understanding of the methodology of fundamental physics. In reply, one should first note that the asymmetry between physics and the other sciences in point of empirical irresolubility may be less pronounced than claimed: Miller points to the theory of "spontaneous generation" as a one-time alternative in bacteriology (ibid. p. 424). Psychology is a fertile ground for candidates for empirically equivalent theories; Kukla discusses "intentional psychology" with and without an ontology of genuine beliefs, wishes, etc. (see Kukla, 1998, pp. 69f). These cases exemplify weak, not strong underdetermination.

On the other hand, those cosmological and micro-physical theories that have known alternatives do have a very special status: they aim to describe all matter and motion, space and time at a fundamental level, at the very root of what there is. If *those* theories do have genuine empirical equivalents, then this fact has potentially great significance for our *ability to understand* the natural world. Physics evidently has a lead over other scientific disciplines in precision, coherence, scope, lawhood and depth of its explanations. Yet, this by itself does not account for the perceived asymmetry in point of underdetermination. These characteristics are generally considered as virtues in every theory and they come in degrees and in various combinations. No principled epistemic distinction can be built on differences between theories like these. Indeed, the progress and increased interconnectedness of the sciences (witness the rise of mathematical, computational and physical methods in the bio-sciences) makes it likely that whatever observational 'irresolubility' affects theories of physics may in the future affect those in other (natural) sciences as well.

There is another objection to consider. The factual asymmetry of cases of underdetermination to which the objection appeals does not say much about the conceptual *possibility* of empirically irresolvable conflicts of assertion in areas of science other than physics. Richard W. Miller, for instance, states that if "smart people" have looked hard over a long time span for an observationally equivalent or better alternative in a given field of inquiry, and failed to find one, then this lack of success is a compelling reason to believe there is no such alternative (Miller, 1987, p. 426). This is, however, not the only and surely not the best explanation. The guiding analogy of Miller's objection is deceiving. Searching the "space of theories" is a different kind of enterprise than searching the garden for a hidden treasure or a checkmate in five moves (S. Leeds): if an extensive search does not turn up one, then probably there is none. There is a world of difference here. The "space of theories" is not conceptually or operationally bounded, while the scientist's search for an alternative theory is a strongly guided, not an open search. The criteria of success are pre-set; the researcher is looking for a theory with this or that general feature. The research methodology, the kind of concepts and ideas that are considered admissible by her or by the scientific community at a time are bounded; time and other resources are limited. The sheer presence of a strong, successful contender skews the social dynamic of research. Failure to come up with an equivalent alternative or better, under conditions that strong, at the very best suggests that there is no such alternative in a certain segment of the total "space of theories".

Furthermore, Miller's 'explanation' is a hindrance to the progress of science in any area. It stifles progress by signalling the scientist that he or she need not bother to search further and better stick to the orthodox set of ideas. By way of example, the best explanation for why generation after generation of very sharp minds failed to come up with a proof for Fermat's theorem would be that the theorem is wrong or 'unprovable' in some sense. Fortunately, few mathematicians accepted the best explanation, and discovered in a long search interesting lemmata and methods and finally a proof of Fermat's theorem.

The history of science shows in many areas a progressive series of successor theories (incommensurability aside), like the sequence of theories starting with Stahl's phlogiston theory of combustion to Lavoisier's to Dalton's atomic theory to modern quantum chemistry. Each one of which is (at least ideally) compatible with the bulk of the evidence that supported its predecessor. The string of theories and conceptual innovations that superseded Stahl's were *in a sense* "present" in his time, although perhaps conceptually inaccessible to him and his contemporaries. One commentator concludes "that we have, [...] in virtually every scientific field, repeatedly occupied a position in which we could conceive of only one or a few theories that were well-confirmed by the available evidence, while the subsequent history [...] has routinely (if not invariably) further, radically distinct alternatives as well confirmed [...] as those that we were inclined to accept on the strength of that evidence" (Stanford, 2001, p. 9). The underdetermination in these cases is not 'strong', rather it is the weak or "transient" kind of underdetermination. Nevertheless, the historical perspective indicates again that the mere fact that nowadays no one can think of an alternative to orthodox theory X is of limited epistemological or ontological consequence.

Some such considerations may have moved I. Douven to replace the thesis

> Any scientific theory has (indefinitely) many empirically equivalents rivals.

as the locus of the anti-realist argument from observational underdetermination with the thesis

> To any scientific theory there *might* be (indefinitely) many empirically equivalent rivals at which we could actually arrive/have arrived.(Douven, 1996, pp. 67, 79)

Douven argues that this thesis poses no threat to realism since it is very likely false. Consider the sceptical hypothesis 'There might be unicorns', which might be taken to throw doubt on current taxonomy. However, if there were any unicorns (on earth) scientists would very likely have found by now some evidence for their existence. They have not, so the sceptical hypothesis is likely false. Similarly, if there were 'accessible' alternatives to any scientific theory, the reason goes, scientists would very likely have discovered them by now. We have not, hence this sceptical thesis is likely false (Douven, 1996, p. 81).

Quite plausibly (even trivially) the second thesis is false, namely when the phrase 'accessible alternative' is interpreted in a suitable narrow way. If an 'accessible alternative' to a given orthodox theory is one, which is required to share a number of significant structural features with the latter, then it may be strictly provable that the space of theories does not contain such an alternative in the proximity of the given theory. A local result says nothing about the global existence of an alternative under a more liberal notion of formal accessibility. Thus much depends on a precise and sufficiently non-*ad hoc* definition of the phrase 'accessible alternative'. I am sympathetic to Douven's intent of using this notion to curtail arbitrary model-theoretic re–interpretations of a given theory (as a mechanism for generating alternatives). Yet, in view of Stanford's point about the retrospective existence of alternatives, the realist would need to develop a psychological-historical account of conceptual (in-)accessibility. The realist also needs to provide an epistemological argument to the effect that conceptual *in*accessibility in the psychological-historical sense matters for a justification. On both questions no detailed answers are available to my knowledge, and the prospects are not very good.[18]

The final worry I turn to is that the observational underdetermination of theory is a manifestation of the familiar problem of *induction*. To tackle the one means tackling the other; weak and strong underdetermination are of the same ilk. If the problem of induction has a solution, then so has the problem of observationally irresolvable conflicts (Giere, 1993, p. 4). Putnam, for instance (Putnam, 1972, p. 360), parallelizes the task of ranking hypotheses in the context of fitting a curve to a set of data and the task of grounding a *rational* ranking of empirically equivalent theories. The two problems have characteristics in common but these are out-weighted by significant differences. Both kinds of problems arise because our hypotheses and theories tend to go beyond any set of observations; in both a criterion of *simplicity* has frequently been evoked to warrant one alternative hypothesis over others. I will now attempt to disentangle the issues.

Suppose, first, that sceptical problem of induction is solved in the sense that a way has been found to justify or prove the one thesis, which Hume thought could alone (potentially) justify inductive reasoning from the observed to the unobserved in a rational manner: the 'uniformity of the course of nature'. Consequently, a rational individ-

---

[18] J. S. Mill sought to explain the dearth of examples of underdetermined alternatives in certain areas of inquiry by the psychological mechanisms of concept formation: the mental process of abstraction from things or events actually experienced by an individual. Mill's tentative explanation has many defects, yet it is an intriguing suggestion for investigating the contingencies and asymmetries of underdetermination.

ual is entitled, subject to certain constraints, to make inferences from an observed sample to a universal hypothesis or a future event. The problem of choosing the 'true one' among underdetermined, conflicting (total) theories, on the other hand, is entirely unaffected by the truth or falsity of the uniformity thesis. The latter problem arises when two or more alternatives make the same predictions (without being notional variants), or in an ideal situation, where two or more (total) theories are empirically faultlessly adequate (whether we are aware of this happy circumstance or not). Under these assumptions, whatever is meant precisely by 'uniformity of the course of nature' (perhaps the existence of stable natural laws), the uniformity thesis and hence the hypothetical solution to Hume's problem cannot make a difference for the problem of rationally choosing between empirical irresolvable, conflicting (total) theories. For the course of nature is uniform under all alternative descriptions. In addition, it seems that Hume was concerned about the empirical adequacy of the conclusion of an inductive inference, not its truth. The alternative theories considered here do not differ in point of empirical adequacy.

It is instructive to turn for a moment to Reichenbach's proposal for a solution of the problem of induction. For that purpose let us distinguish two types of inductive problems. Type (i) is the problem to justify the use of a given inductive rule or account of confirmation. Type (ii) is the problem to justify a criterion for selecting those hypotheses to which the account can be applied, if it can be applied to all. The latter problem was emphazised by N. Goodman as the "new riddle of induction"; I'll turn to its significance for the problem of underdetermination below. Reichenbach famously proposed a "pragmatic" solution to the first problem (Reichenbach, 1938, pp. 348–349). In the case of induction by enumeration, provided there actually is a fixed ratio of As among all observed Bs, a rule exists, which can be shown to converge in the limit to the truth: conjecture that the ratio of observed As among observed Bs is equal to the hypothetical limit-frequency and up-date the relative frequency with each successively larger sample if necessary. Similarly, the iterated choice of the "simplest" interpolation in the curve-fitting problem, i.e. a chain of straight-line segments connecting adjunct pairs of ideal, error-free data-points, will in the long run converge to the true curve, provided the limit exists at all (Reichenbach, 1938, pp. 377–378).[19] Thus there is no need to establish the "uniformity of nature" *a priori*; if the grand principle is false no inductive method (including fortune-telling) works, and if true the standard inductive rules are truth-conducive.

Now, one objection to the "pragmatic" solution is that it is conditional on nature's cooperation of which we remain ignorant. But a more telling objection is that there are many other inductive rules beside the orthodox ones, which can be justified in the same manner, but make markedly different predictions in the short run. Uncountable many others beside the "simplest" curve (straight-line segments) will *ceteris paribus* converge to the true curve as well. (In addition, simplicity and goodness-of-fit in the curve-fitting problem tend to pull in opposing directions, as Philip Frank pointed out.) Thus alternative inductive rules are themselves (weakly) underdetermined by past data. Which rule is one to trust? Whatever the merits of Reichenbach's account in this respect, it does not extend easily to the case of rational choice between empirically (hypothetico-deductive) equivalent theories or total theories, since the crucial possibility of continuous empirical self-correction is lacking. All relevant empirical information, it is assumed, is taken in already. Hence Reichenbach's distinction between an inductive rule based on *inductive simplicity*, which has predictive value, and a methodological rule based on *descriptive simplicity*, which has no empirical consequences and (for that reason) is of only heuristic value instead of epistemic value (Reichenbach, 1938, pp. 373–374). This bears on Putnam's suggestion of transferring the truth-conducive principle of simplicity from the class of curve-fitting problems to cases of full observational indistinguishability or strong underdetermination. The condition for its applicability in the former case is that the results of selections guided by the principle can be monitored, if only on an infinite sequence of data. This condition is not satisfied in the very cases, which are of interest here. To apply the principle in an entirely

---

[19]Compare the argument in Sober (1996) that relies on a statistical theorem by H. Akaike, where simplicity is measured by or correlated with the number of free parameters in the model, as opposed to an algorithmic notion of simplicity, compare Glymours (1980b, pp. 309, 331), or Reichenbach's line segments.

different context and to maintain that it is more than a heuristic is *ad hoc* and arbitrary. Reichenbach was surely right in drawing a sharp distinction between inductive simplicity and *descriptive simplicity*, a distinction that robs the methodological similarity between Humean induction and strong underdetermination of its significance.

There is another twist to consider. Arguably, it is not long run certainty inductive agents are after but certainty or high probability in the short run. The principled distinction between the rational choice problem that the underdetermination thesis poses and inductive reasoning, type (i), breaks down if the justification of short run practices of acceptance or belief is the core of the problem. From this point of view, empirical underdetermination and induction are instances of the same general problem. Observational irresolubility in theories and the problem of induction (i) are aspects of a "single problem of *nondeductive* inference" (Friedman, 1983, pp. 273). However, in 'choosing' (rationally) between empirically equivalent theories there is no short and long run, measured as progressive confirmation or disconfirmation along a sequence of fresh observations and new experiments. The alternatives stand and fall together (I bracket the possibility of switching auxiliary hypotheses here), and are – in one version – consistent with all observations. In orthodox cases of inductive reasoning, by enumeration or even by elimination, the (ideal) scientist settles on a hypothesis when he thinks he has sufficient evidence at his disposal and serious alternatives are eliminated (apart from other constraints and goals he may have to optimize). But this kind of reasoning cannot be exactly the one that governs the short run choice between strongly underdetermined alternatives. It is quite possible that one of these alternatives somehow "meshes" better with the data than an hd-equivalent alternative. A scientist may come to realise this logical fact and judge accordingly in the short run, yet this is akin to logical learning not ordinary inductive reasoning.

Choice among strongly underdetermined alternatives is more closely related to the second type of inductive inference, (ii) above.

Indeed, J. Earman suggested that Goodman's riddle is a version of the (older, well-known) curve-fitting problem, or rather both are instances of the universal phenomenon of weak underdetermination, of more than one hypothesis fitting any given finite string of data (Earman, 1992, p. 110). The question of how the scientist is to deal with this embarrassment of riches and the further problem of rationally justifying a choice among strongly underdetermined alternatives theories are related. A justified, objective criterion for how to limit (for the purpose of testing and confirmation) the super-abundance of alternatives to a few "serious" contenders in the first problem would suggest a corresponding criterion for the second. For instance, if the Goodmanian "anchoring" of hypotheses is the criterion and condition for straight inductive inferences, why not select among strongly underdetermined alternatives accordingly, i.e. with an eye toward the predicates employed and their prehistory in successful scientific practice for each alternative? If hypotheses are to be ranked and selected according to their *a priori* probability, as assigned by an individual scientist in the light of his background information, then why not transfer the approach to strongly underdetermined alternatives? These questions have negative answers, I believe, but the point here is to emphazise the relationship between the two problems.

1.5 Three Responses and Strategies

I distinguish three main families of analyses of empirically irresolvable conflict. They shape the course of my argument, as I will attempt to indicate the limits of each of them.

(1) The first type of analysis contends that the appearance of strongly observationally 'irresolvable', conflicting statements is due to a defective account of justification and methodology (for instance, Quine's deductivism). Every empirically irresolvable conflict is contingent on a specific account of justification. Faced with an irresolvable conflict of assertions in the sciences it is natural to think that the error is in the presumed canon of justification. A strengthened and sophisticated account of how theoretical statements are confirmed on the basis of relevant experience will surely reveal that one statement

1.5. Three Responses and Strategies

is better supported by the data than the other. Reflection on proper methodology, or causality, will remove all or most alleged cases of underdetermination of theories of wide scope (Richard Boyd, Richard I. Miller, Clark Glymour among others). This response actually dates back to Descartes and is currently the most popular and perhaps most promising strategy among the three to be examined.

As I have pointed out above one influential thinker whose epistemological and metaphysical views rely on a specific account of justification, is Quine. This account of justification puts the implication between sentences at its center, particular those implications that terminate in observation sentences. It is a version of classical *hypothetico-deductivism*. Despite its strong credentials orthodox deductivism (in its many variants) has become the target of much criticism. Alternative accounts of confirmation and justification have been proposed over the last decades, and a few will be inspected in a separate chapter, but none can command anything close to a consensus. Chapter 3 attempts to evaluate the prospects of the methodological response to the issues raised by the thesis of strong underdetermination.

(2) The second analysis to be discussed has been unduly neglected recently, I think. This analysis is based on the idea that the origin of observational underdetermination of theories lies entirely with our conceptual representation of the facts, i.e. in the way we represent the world in our theorizing.[20] The terms and expressions of those conflicting statements, strongly underdetermined by 'all possible' data, contain by necessity non-observational and theoretical expressions. It is plausible to conclude that one of these expressions lacks proper reference, or is 'empty', rendering the statements 'meaningless', or that the truth-values of statements essentially involving such expressions can be settled by way of convention. The dispute over which of the conflicting alternatives is true is, on this account, *purely verbal*.

That an empirically irresolvable conflict of assertions licenses an inference to semantic deficiencies in the assertions is a theme that permeates 20th-century philosophy. Schematically, the eliminative argument

---

[20]It has influential proponents in H. Reichenbach, P. Horwich, H. Field among others.

proceeds by isolating a specific claim, say F, and constructing an empirically (inferentially) irresolvable alternative that lacks the very feature expressed by F, or has consequences incompatible with F. (Sometimes, actual construction is replaced by plausibility arguments for the *constructibility* (or logical possibility) of a conflicting alternative, by noting that nothing, no fact, no physical law, etc., forbids the construction.) A comment by Richard M. Hare on the objectivist–subjectivist debate in value ethics furnishes an example of the eliminative use of irresolvable conflicts. He argues against the very *intelligibility* of the objectivist position in the following way:

> Think of one world into whose fabric values are objectively built; and think of another in which those values have been annihilated. And remember that in both worlds the people in them go on being concerned about the same things [...] Now I ask, "What is the difference between the states of affairs in these two worlds?" Can any answer be given except "None whatever"?[21]

Logic only dictates that the existential assumption of objective values is strictly irrelevant, or surplus, for the purpose of explaining or guiding action. Hare's conclusion, that the objective existence of ethical values is *unintelligible*, can be reached from one of two additional premises that differ in strength. The *verificationist* version of the argument from inferential indistinguishability invokes an explicit criterion of cognitive sentence meaning. To paraphrase Ayer, a declarative statement is cognitively meaningful, if and only if directly testable experiential propositions can be deduced from it that cannot be deduced without it. Meaningful statements in a given language are thus associated (individually) with a set of empirical circumstances that count as refutation or verification. The existence of an observationally irresolvable alternative, together with the verification criterion, entails that any statement involving F lacks cognitive meaning.[22]

---

[21] Quoted from J. L. Mackie (1977). *Ethics*. London: Penguin, p. 21.
[22] Empirical irresolubility does not necessarily lead to reduction and *elimination* of a surplus feature F, although this tends to be the most frequent conclusion.

1.5. Three Responses and Strategies

Carnap, who first took on irresolvable conflicts in *Scheinprobleme in der Philosophie*, gave a classical expression of the verificationist resolution of observationally un-resolvable conflicts of assertion. Beside a discussion of the problem of other minds, Carnap addressed the competing ontological claims of "naive" realism on the one hand and "idealism", or anti-realism in general, on the other, in the light of phenomenalism. It is the impression of un-decidability between these rival interpretations of the totality of our experiences, both by empirical means and by rational argument, and hence the dogmatism involved, which motivates the resort to the issue of sentence meaning. A sufficient and necessary requirement for declarative statements to have meaning is:

> bringt eine (vermeintliche) Aussage keinen (denkbaren) Sachverhalt zum Ausdruck, so hat sie keinen Sinn [...] Bringt eine Aussage einen Sachverhalt zum Ausdruck, so ist sie jedenfalls sinnvoll.

Consequently, conflicting statements, if observationally indistinguishable, do not have cognitive meaning.

Reichenbach seconded Carnap when he claimed that whatever criterion of cognitive meaning is imposed on a discourse the justification for doing so is a conventional methodological decision. The specific progress Reichenbach's works represent for our problem is his claim:

CO All contradictions between theoretical statements of observationally equivalent descriptions of the facts can be traced back to the

---

The *Turing Test* is taken to support a decidedly non-reductive conclusion in the research program of Artificial Intelligence. Turing's empirical criterion states that if any human judge is unable to tell from extended (ideal) verbal tests whether an interview partner or pen pal, whom she cannot visually "check out", is a human or a program (robot) then the program has human intelligence. *Turing indistinguishability* (S. Harband) does not support a reductive conclusion – to the contrary, empirical irresolubility is used to infer that the machine *has* a mind. In general, if one of two empirically irresolvable alternatives has the feature F and the other does not, then one may conclude that latter must have the feature, if only disguised in mathematical garb. This line of argument is the basis of Robert Wilson's critique of the "double standard", see chapter 4.

adoption of different *coordinative definitions* (or correspondence rules).

If one discovers a case of observational irresolubility between theories, a proper reconstruction of the theoretical edifice will show that one has failed to strictly fix the cognitive meaning of all terms. Only relative to a determination of meaning, through selected relations to observables, the contradictory claims have meaning at all. Conversely, truth values can be assigned at will over empirically irresolvable assertions (as long as consistency is saved) since they are presumably disconnected from even the possibility of empirical support. Such an assignment amounts indirectly to the adoption of *coordinative definitions*. Empirically irresolvable conflicts of assertions are *litmus* tests for the status of conventionality in statements (to use a phrase of Grünbaum's). Reichenbach thus proposed a *prima facie* plausible diagnosis of the phenomenon of empirical underdetermination in the exact sciences. Moreover, his account of the role of coordinative definitions offers an *explanation* of why, inevitably, alternative but empirically equivalent conceptual frameworks, or "equivalent descriptions" exist. I will return to this important issue in chapter 4.

Attractive as the semantic diagnosis of empirically irresolvable conflicts must appear, there are three cardinal obstacles. (a) There is no criterion of meaning, which is not either too wide or too narrow, i.e. either permits all consistent statements as meaningful or excludes even good scientific hypotheses. (b) Verificationist criteria are incompatible with holism of confirmation, and the latter is a well-argued doctrine since the times of Duhem. (c) Any criterion of meaning, as a premise for the resolution of irresolvable conflicts, is in need of compelling independent justification. The main reason cited for adopting this or that criterion is that it conforms with, and even is characteristic of, the natural sciences. Yet, (a) shows that this claim is far from true, and as a factual statement about scientific practice it is of debatable epistemological import. Actually, Reichenbach's conventionalist view of criteria of meaning is quite a defense of verificationism, since it dispenses with the objection (viz. Putnam, Barry Stroud) that the verification criterion itself has not been properly justified, or that it is less well founded

1.5. Three Responses and Strategies

than the doctrines it was used to undermine. The reply is that these criteria neither need nor are capable of a foundation.

Our choice among them is guided by pragmatic concerns and conservativeness alone. The concept of truth by convention, however, has been very effectively criticised in the companion notion of analyticity (at least for logical and mathematical truths). Consequently, conventionalism, including the Reichenbachian variant, lost its appeal.[23] Nevertheless, a dominant motive for adopting verificationism in the first place was its supposed efficiency in resolving irresolvable conflicts. Given that the explication of sentence meaning is the thorny – if not misconceived – problem that it is, note that there are less all-encompassing, defendable approaches to meaning tailored to the problems posed by strong observational underdetermination or empirical irresolvability. The naive verificationist inference to the literal meaninglessness, or unintelligibility, of statements that together make up an empirically irresolvable conflict is stronger than necessary. I discuss sufficient criteria for the lack of cognitive significance or content in section 5.4.

I turn to the last and most radical approach within this family. The proposal is to reject the *principle of bivalence* for empirically (rationally) irresolvable statements, and either (a) to deny outright that empirically irresolvable statements have an *objective truth value* or (b) claim that their respective truth value is objectively indeterminate. In both cases the statements are regarded as perfectly cognitively meaningful, something that proponents of a verificationist semantics reject. I include this proposal in the present section, although linguistic meaning is not my focal issue, since it aims to show that empirical irresolubility is a sure sign of an empty and "verbal" dispute. Quine considered and ultimately objected to the rejection of bivalence, but Hartry Field has recently advocated option (b) for the case of rational irresolvability in mathematics, i.e. "metaphysical irrealism" in mathematics.[24]

---

[23]I suspect that it is true that much of the conventionalist program, though not the bridge-law approach to cognitive meaning of theoretical expressions, can be recovered by substituting 'conventional truth' by the toned-down concept of "content-neutrality" with respect to the facts under investigation.

[24]For Quine's view see chapter 3 of Quine (1981).

The proposal involves non-standard probabilities or, alternatively, a multi-valued logic. I consider Field's views in section 6.4.

(3) The last one to be considered is Quine's own response to the underdetermination thesis on the level of 'total science'. As was pointed out, Quine wanted to read the thesis *not* as a 'theoretical' claim, but as a temporary, contingent record about success in translation – translation between our 'global science' and a hypothetical rival system of belief, which is nevertheless observationally equivalent, see section 6.5. It is helpful to be clear about Quine's bifurcated use of empirical irresolubility: topic-specific and with regard to "global science". On the one hand, he harnessed empirical irresolubility specifically to undermine what he called "mentalistic" theories of meaning. On the other hand, Quine aimed to restrict the possibility that irresolubility affects our "global system of the world", i.e. the authority of the sciences to explain the world, relying on a technical idea due to Donald Davidson, and a 'pragmatic re-interpretation' of underdetermination.

Chapter 2
Underdetermination Issues in the Exact Sciences

This chapter is divided in two parts. The first part looks at basic logico-mathematical notions and theorems, which mark the cornerstones of the philosophical debate that is the subject of my essay. The second part examines a few typical examples of "equivalence" in physical systems and underdetermination of physical theories. These are the facts; they call for an interpretation.

2.1 Logical Equivalence, Interdefinability, and Isomorphism

The empirical content of a theory is the class of its consequences that can be stated in the observation (sub)language chosen. Its *relative* content is the observational consequence class derivable from the theory together with a set of auxiliary hypotheses (see section 3.1), but for the following considerations the distinction does not matter much. If the empirical content of two theories is exactly the same, it seems reasonable to assume that the theories differ only as *formulations* or "notational variants" do. Indeed, a theory may be identified with the equivalence class of all "notational variants". One writer in the Carnapian tradition (Rozeboom, 1960, p. 373) claimed accordingly that

> Two theories are equivalent if and only if they have the same observational consequences.

Logical equivalence, interdefinability and isomorphism are the main candidates for identity criteria for theories, a key concept for evaluating underdetermination claims. I begin by recalling these familiar concepts.[1]

Logical equivalence. The sentences "Sophocles is human" and "Sophocles is a featherless biped" have the same truth value. They are *materially equivalent*, but not logically equivalent: there is no logical necessity in them having the same truthvalues. They differ in *meaning* and are not synonymous. Logically equivalent sentences are for instance "Sophocles = Sophocles" and "Sophocles is bald = Sophocles is bald". The difference is captured as follows. The formulas ϕ and ψ of a given (countable) first-order language L are *"logically equivalent"* if for every model (world) in which ϕ is true ψ is true and vice versa. Given this definition, two formulas ϕ and ψ in L are logically equivalent if and only if the biconditional $\phi \leftrightarrow \psi$ is valid (i.e. true under every interpretation). Any two valid formulas are logically equivalent. If two formulas ϕ and ψ are logically equivalent, one can generate from an arbitrary formula τ that contains the formula ϕ a logically equivalent one, by replacing ϕ by the formula ψ. Sometimes it is useful to relativize and restrict logical equivalence of sentences to a given theory: we say that ϕ is equivalent to ψ *modulo the first-order theory T* if for every model M of T ϕ is true in M if and only if ψ is. By way of generalization, two first-order *theories* in L are said to be logically equivalent if they have exactly the same models. Intuitively speaking, equivalent theories share the same non-logical vocabulary and have the same theorems and differ only in the way they are axiomatized. Two first-order theories in L are equivalent if and only if one is an *extension* of the other. Note finally, it can be shown that for every expression in first-order predicate calculus there is an equivalent one – equivalent in point of both being valid in all non-empty domains – that contains only *one two-place predicate* (Hilbert and Ackermann, 1959, p. 124). Fixing the interpretation of this one predicate does it for the theory.

The logical concept of *variant* helps to cut down on formulas that differ only in notation through the choice of variables. One formula is said to be a *variant* of another in L if each can be got from the other by a consistent replacement of variables. For instance, $\forall y \phi(x, y)$ and

---

[1] This is the stuff of many textbooks, I found Hodges (1997) particularly helpful.

$\forall z \phi(x, z)$ are variants. The relation "variant" is an equivalence relation on the class of formulas (Hodges, 1997, p. 26). A set of sentences S_2 (in a first-order language with a finite number of predicates $P_2, R_2, \ldots$) is a *"notational variant"* of the set S_1 if and only if it is logically equivalent to the set generated from S_1 by replacing the n-place predicate P_1 of S_1 by the n-ary predicate P_2; R_1 of S_1 by R_2, etc. (Terms like functions and constants in a theory can be defined in terms of relations in the customary way.) In *Logische Syntax der Sprache* Carnap used the notion of variance to characterize what he calls an isomorphism of languages: "Angenommen zwei Sprachen S_1 und S_2 verwenden ungleiche Zeichen, aber so, daß sich eine eineindeutige Zuordnung zwischen den Zeichen von S_1 und denen von S_2 herstellen läßt derart, daß jede syntaktische Bestimmung in bezug auf S_1 in eine solche in bezug auf S_2 übergeht, wenn wir sie anstatt auf die Zeichen von S_1 auf die jeweils zugeordneten Zeichen von S_2 beziehen, und umgekehrt. Dann sind die beiden Sprachen nicht gleich, aber sie haben dieselbe formale Struktur (wir nennen sie isomorphe Sprachen)" (Carnap, 1963c, pp. 5–6). Carnap relativized "equivalence" to "L-equivalence" with respect to the underlying formal language L. Whether this notion of logical equivalence guarantees "equivalence of meaning" is a bone of contention. Do they have the *same* meaning? Wittgenstein maintained in *Tractatus* (5.141) that sentences that mutually imply each other say the same thing, a thesis that N. R. Hanson elaborated with regard to the "empirical equivalence" (or weak underdetermination) of Copernican and Ptolemaic terrestrial astronomy Hanson (1966). There is no need to examine this interesting debate, since questions of synonymy or "intensional isomorphism" (Carnap) play no role in contemporary discussions of the underdetermination thesis.

The notion of logical equivalence is *too strong* in many respects. Accordingly, discussions of underdetermination and identity criteria of theories turn to the idea of *definitional equivalence* between theories. A theory T_2 in a first-order language $\mathcal{L}_2$ and a theory T_1 in a first-order language $\mathcal{L}_1$ are *"definitional equivalent"* if and only if explicit definitions U_1 of the symbols of $\mathcal{L}_2$ in terms of $\mathcal{L}_1$ can be found, as well as explicit definitions U_2 of the symbols of $\mathcal{L}_1$ in terms of $\mathcal{L}_2$, such that $T_1 \cup U_1$ is logically equivalent to $T_2 \cup U_2$. When two theories are definitionally equivalent in this sense, one can expand a model of T_1 into a model of T_2, and vice versa (Hodges, 1997, pp. 54–55). There are limits to the usefulness of this notion too since it is syntactical in nature. Mineralogy and zoology could turn out be strictly interdefinable when properly formalized despite their different subject matter. A number of philosophers have adopted this notion as a necessary, not sufficient criterion for the identity of alternative theories. Newton-Smith writes:

> two theories T_1 and T_2 are notational variants if they have definitional extensions T_1' and T_2', such that any theorem of T_1' is a theorem of T_2' and vice versa.[2] (Newton-Smith, 1978, p. 97)

Putnam is perhaps the earliest source with regard to philosophical applications: "The definition of T_1 as relatively interpretable in T_2 is: there exists possible definitions [...] of the terms of T_1 in the language of T_2 with the property that, if we 'translate' the sentences of T_1 into the language of T_2 by means of those definitions, then all theorems of T_1 become theorems of T_2. Two theories are mutually relatively interpretable if each is relatively interpretable in the other [...]" (Putnam, 1979, p. 38).

From here it is a small step to Quine's identity criterion for underdetermined alternatives:

> two formulations express the same theory if they are empirically equivalent and there is a reconstrual of predicates that transforms the one theory into a logical equivalent of the other. (Quine, 1975, p. 320)

A "reconstrual of predicates" is a permutation of the predicates of the underlying language; more precisely, a rewriting of predicates into corresponding open sentences followed by a permutation. For Quine logical equivalence between two theories is an unambiguous notion only in

---

[2]Compare on this Putnam (1975a, pp. 38–39), ibid. p. 121, fn. 15; Wilson (1981, p. 41) and Friedman (1983, pp. 281–283, 285). Similarly, Paul Horwich defines: "Let us say T_1 and T_2 are *isomorphic* just in case there are two formulations of them which are potential notational variants of one another" (Horwich, 1982, p. 65).

"our" first-order logic as *the* scientific language. For theoretical alternatives to a given theory in other languages the problem of translation interferes, and the notion of implication has to be relativized (Quine, 1975, p. 318).

Isomorphism. It is suitable to begin with the basic concept of the equivalence between two *sets*. Sets A and B are called "equivalent" or "equinumerous" ($A \sim B$) if there is a one–one correspondence between the elements of A and the elements of B. It is easily verified that $\sim$ is a reflexive, symmetric and transitive relation on sets. An initially baffling consequence of this familiar definition is that for infinite sets both $A \sim B$ and $A \subset B$, or A is a proper subset of B, can be true.[3] The equivalence of sets thus defined is *coarse-grained* in one important respect: it ignores differences in the ordering superimposed on the sets A and B. I turn now to the concept of isomorphism between models or structures, which generalizes and fine-tunes the set-theoretic notion of equivalence, by taking differences in the respective ordering into account. Recall, that a relation R is a "partial ordering" in the set A if it is reflexive, anti-symmetric and transitive in A. If such a relation exists for all elements of A R is called a linear (simple) ordering of A. In other words, for every pair of elements of A there is a rule which determines which of the two is "larger than" (precedes, smaller than) the other. The ordering relation is conventionally symbolized by $\leq$. Loosely speaking, an isomorphism between two partially ordered sets $\langle A, \leq \rangle$ and $\langle A', \leq' \rangle$ is a one-to-one correspondence π between the sets A and A' such that both π and its inverse π^{-1} preserve order. If such

---

[3]The notion of "one-to-one correspondence" can be dispensed with in the definition of set-theoretic equivalence, as the following consideration shows. Suppose A and B do not share a common element. Consider the set of all element-pairs $\{a, b\}$, itself an element of the *product set* $A \times B$, such that (i) $a \in A$ and $b \in B$; (ii) two different element pairs never have same element from A or from B. If a required set of pairs exist, a one-to-one correspondence between A and B exists, and if there is no such set of pairs, no one-to-one correspondence can exist. The refined definition of set-theoretic equivalence ("similarity") may now be stated as follows: two sets A and B, which do *not have a common element*, are "equivalent" if the product set $A \times B$ has a subset Π, such that every element of $A \cup B$ appears in exactly one element of Π. The assumption of the sets having no element in common is crucial for the formulation, but is not essential.

a correspondence exists, the two ordered sets are *isomorphic*. Isomorphism like equivalence between sets, discussed above, is a symmetric, reflexive and transitive relation on the class of ordered sets. A simple consequence is that if A and A' are finite, then $A \sim A'$ if and only if $\langle A, \leq \rangle$ and $\langle A', \leq' \rangle$ are isomorphic.

The linearly ordered set can be turned into an example of a *structure*: an object composed of a domain (a set), the universe of discourse; a number of "concrete" n-ary relation and function symbols on the domain; together with a number of constants of the domain. The given, concrete relations, etc. are called the signature of the structure. For instance, the reals with addition and multiplication form a structure: $\langle R, +, -, \times, 0, 1, \leq \rangle$.

Definition 2.1.1 *Two structures A, B with signature L are called isomorphic, if and only if there is a mapping π such that: (1) π is a bijection of the domains of A and B; (2) for each constant c of L $\pi(c^A) = c^B$; (3) for each n-ary relation R of L and n-tuple a from the domain of A, if $a \in R^A$ then $\pi(a) \in R^B$; (4) for each n-ary function f of L and each n-tuple a from the domain of A, $\pi(f^A(a)) = f^B(\pi(a))$.*

For example, the structure which has as domain the natural numbers, as constant the "0" and as binary relation "+" is isomorphic to the structure which has as domain the even natural numbers with addition and the "0" as a constant. The *Isomorphism Lemma* states that if two structures I and J are isomorphic then for every the sentence ϕ: $I \vDash \phi$ if and only if $J \vDash \phi$. Isomorphism between two (physical) theories is too fine-grained a notion for our purposes. Intuitively, the natural numbers with addition and multiplication exemplify the same structure as the natural numbers with addition, multiplication and less-than. They are not isomorphic. However, the binary relation less-than $(x < y)$ can be defined explicitly in terms of the first system as: $\exists z(z \neq 0 \,\&\, x + z = y)$ (Shahan and Merrill, 1977, p. 91). Similarly, there are various formulations of Euclidean plane geometry that have different primitives. In the standard first-order background language the natural numbers with successor alone are not equivalent to the natural numbers with addition and multiplication, because addition cannot be defined from successor in a first-order language. The theories are equivalent in a second-order background language. Hence, equivalence depends on the "background language". M. Resnik defined the concept of a "full subsystem" to circumvent these difficulties: P is a full subsystem of R if P and R have the same objects and if every relation of R can be defined in terms of relations of P. Two systems M, N are *structure equivalent* if there is a system R such that M and N are isomorphic to a full subsystem of R. The notion of isomorphism between two structures reflects intrinsic properties of them, being defined without appeal to a language, whereas the relation (logical) equivalence involves the language. However, it is not true that structures in which the same sentences are satisfied need to be isomorphic.

The basic notions defined above have been generalized in many ways: from logical equivalent to "elementary equivalent" and "back-and-forth equivalence"; from isomorphism between interpretations to "partial isomorphism" and "finitely isomorphic". For the most parts, however, these fruitful developments do not throw light on the issues examined in the present monograph.

The following theorem states that under certain assumptions two empirically equivalent theories have *no models in common*; their intended interpretations differ necessarily (Glymour, 1971, pp. 282, 286). A fragment of Euclidean geometry, as formulated in first-order logic provides an illustration. The theory has an "empirically equivalent" alternative in a (first-order) fragment of elliptic geometry. They have no models in common.

Theorem 2.1.1 *Let L be a first-order language with identity and a finite number of predicates, and suppose T is a theory in L such that*
(i) every complete and consistent extension of T is decidable;
(ii) the collection of observational consequences of T has at most a finite number of finite models (up to isomorphism).

Then there is a finitely axiomatizable theory T^* which is observationally equivalent to T and which has no models in common with T^*.

The philosophical significance of the theorem is limited. Due to assumption (ii) there are no interesting examples known in the empirical sciences. As an illustration for the severity of (ii), note that sentences that admit only finite models tend to be not very interesting ones, like: $\forall x \forall y\, x \equiv y$. Despite its limited applicability, Glymour's theorem puts the claim that empirically indistinguishable theories *must* say the same thing, have to be 'identical', in perspective.

2.2 Theorems of Ramsey and Craig

Craig's much-discussed theorem appears to *undermine* the underdetermination thesis. Craig's theorem demonstrates that under certain mild assumptions theoretical concepts can be "eliminated" in the sense that there is a romp theory that codes all and only the observation sentences of the original theory. Surprisingly, the romp theory is not only equivalent to the empirical content of the original theory, but *is a theory* in that it is recursively axiomatizable if the original theory is recursively enumerable, and expressible within the underlying logic. One might have expected, for example, that new and complicated *rules of inference* are needed to compensate for the theoretical superstructure.[4] All this is water on the mills of a view of theories as symbolic devices for correlating patterns of observations (Craig speaks of "auxiliary expressions"). Theories are individuated only by their empirical content. Nothing else but an observation sentence in the consequence class with a different truth value can count as a genuine difference between alternatives. In sum, by giving a boost to the eliminativist (instrumentalist) philosophy of science, Craig's theorem underwrites a strong identity criterion for theories that rules out the underdetermination thesis.[5]

Theorem 2.2.1 (Craig) *Let T be a recursively enumerable theory, and consider any division of the terms of T (finite in number) into two disjoint sets: V_O and V_T, the observational and the theoretical vocabulary respectively. Let T_O consist of those theorems of T which*

---

[4]Hans Hermes has investigated the intriguing possibility of replacing theoretical terms by non-standard inference rules operating on observation sentences Hermes (1950).

[5]The best introduction to the theorem and its philosophical context is still Craig (1953).

contain only terms from V_O. Then T_O is a recursively axiomatizable theory.

The formulation employs standard terminology with the exception perhaps of the notion "recursively enumerable" set.[6] The term denotes sets (e.g. of axioms or numbers) whose elements can be mechanically generated if the procedure is allowed to continue indefinitely, while no procedure exists that always determines whether or not a given number or axiom belongs to the set. The set of theorems of first-order logic is such a (non-recursive) set. It is not necessary for us to review the proof of the theorem. An illustration brings out its essence English (1973). The toy theory T is manifestly recursively axiomatized with just the two axioms:

$$F$$
$$F \to O$$

where F is the one and only theoretical and O the observation sentence. The empirical hd-content of T is given by one sentence O. The secondary theory T_O has for axioms a specific subset (T^C, say) of all finite repetitive conjunctions of the form $O\&O\&\ldots O$. This subset can be recursively specified by the rule: A formula is an axiom of T^C if and only if the number of its conjuncts O is equal to the Gödel number of a proof of O in the system T. Since there are innumerable (trivial) ways of proving O from T (for instance from the four premises F, F, F and $F \to O$) T^C has an infinite number of axioms. The alternative axiomatization of T's empirical content thus turns out to be a consistent, decidable theory.

Consider the conflicting, *hd-underdetermined* alternative T' with axioms:

$$\neg F$$
$$\neg F \to O.$$

The notation is as above. The corresponding romp theory T'^C can be recursively specified as before. Since none of the proofs of O in T and

---

[6] Adapted from Putnam (1965). Putnam's proof does not depend on the device of Gödel numbering, as does Craig's original proof.

T' respectively can have the same Gödel number, due to the presence of the negation sign, the corresponding romp theories T^C and T'^C will be disjunct (both of course entail the very same sentence O). The moral of the exercise is the following: while the finite theories T and T' are incompatible their infinite Craig replacements T^C and T'^C *are compatible*. So, *if* the essence of theories is fully captured by their respective Craig replacements, then the underdetermination thesis is provably false.

The "if" is a big if. Scientists value the axiomatization of a theory because it reflects fundamental assumptions of a body of knowledge and has great heuristic value. Finitude, organization and conceptual economy are lost in the Craig replacement (as Craig himself emphasized). And since one has to refer back to the original theory when checking which of the repetitive conjunctions belongs to the secondary T^C and which not, full replacement has evidently not been achieved. Accordingly, Quine concluded "Craig's result does not refute the thesis of underdetermination" (Quine, 1975, p. 326). The conclusion is correct, I believe, though it may well be a Phyrronian victory. The mere *psychological* necessity of using finite theory formulations forces a retreat from eliminativism to fictionalism or to the Ramsey view perhaps, but it does nothing to remove worries surrounding the issue of cognitive meaning of theoretical terms and surely does not *justify* the realist interpretation of theoretical terms. On a full-blooded *realistic* view of theories Craig's replacement procedure not only does not refute the thesis, but *proves* a harmless version of the underdetermination thesis: the secondary theory T^C is an account of the relevant facts of its own, alongside and competing with T. Both theories obviously ascribe different ontologies to the world. Craig's procedure then produces a rival, empirically equivalent alternative to any theory T, yet it cannot show that there are unlimited other, potentially more interesting alternatives to T. This is a far cry from establishing the universality thesis (U) (section 1.2).

I now turn to the *Ramsey sentence* associated with a theory. Quine (1995) took the Ramsey sentence as supporting the doctrine of ontological relativity in the case of abstract objects, in that it "defers" choice of the ontology (see section 6.3); there are affinities with

Quine's "proxy-function" construction of alternative 'global systems' of the world (section 4.2). In the Ramsey sentence descriptive theoretical terms are eliminated in favor of second-order variables and additional existential quantifiers, rendering the theory and its empirical content in purely observational and logical vocabulary. The main result, as in the discussion of Craig's theorem, is that the Ramsey sentences of two empirically hd-equivalent, incompatible theories will be *compatible* with each other. Consequently, *if* the Ramsey sentence captures all there is objectively to a theory, than the underdetermination thesis is false. (I refer to un- or pre-ramsified theories as "primary theories", and to the Ramsey sentence and Craig's recursive systematization of the empirical content as "secondary theories".)

Formally, the Ramsey sentence of a primary theory is produced in two steps. Let T be a finitely axiomatized theory with a finite number of predicates, presented as one long single sentence, the conjunction of all its postulates. Suppose the theory in question has only one theoretical term F_1, which we treat as a proper name. The term will be defined with the help of the definite description operator by the expression $F_1 =_{Df} \iota(x)T(x)$. If there are more than one such term present: $F_1 =_{Df} \iota(x_1)(\exists x_2, x_3, \ldots x_n)T(x_1, x_2, \ldots x_n)$. Second step: T's ramsified version is given by $T(F_1, \ldots F_n) =_{Df} (\exists x_1, x_2, \ldots x_n)T(x_1, x_2, \ldots x_n)$. *Logical variables* replace theoretical predicates, and the body of the theory is treated as a *propositional function* with respect to that variable. This function is turned into a testable sentence by prefixing with an existential quantifier with respect to the variable. It turns out that the sentence thus constructed is empirically hd-equivalent to the theory in question. Every standard first-class theory can be ramsified provided the underlying language is a second-order logic. Theoretical terms cannot, however, be eliminated sentence by sentence. In order to strip a simple sentence, like "mass(Jupiter) : mass(Earth) = 318", from the theoretical term mass, the whole of classical mechanics plus its set of coordinate definitions, would have to undergo Ramsification.

By way of illustration, let T be given by the two postulates:

$$(i) \quad (x)F(x)$$
$$(ii) \quad (x)(F(x) \rightarrow O(x))),$$

where O denotes the observational and F denotes the non-observational predicate. The associated Ramsey sentence $R(T)$ is given by:
$$(\exists S)((x)S(x)\&(x)(S(x) \to O(x))).$$
$R(T)$ as an existential generalization cannot have a smaller empirical content as T. That $R(T)$'s content cannot be larger asserts the following lemma:

Theorem 2.2.2 *Let T be an axiomatizable theory, with a division of the terms of T (finite in number) into two disjoint sets: V_O and V_T. Every observation sentence that is consequence of T is a consequence of $R(T)$ as defined above, and vice versa.*

For our purposes important is the following fact: Ramsey sentences with the same set of first order observational consequences are mutually logically consistent. Hence, the underdetermination thesis is false if applied to the class of Ramsey sentences. There is only one Ramsey sentence (up to trivial isomorphisms) that is compatible with the totality of all observation sentences. *In general, underdetermination is false for secondary theories.*

Theorem 2.2.3 *If two Ramsey sentences $R(T_1)$ and $R(T_2)$ have exactly the same observational consequences, then they are compatible.*

The following example shows that the theoretical superstructure of two theories can be incompatible although the theories share the same empirical content.[7] Let T' be given by the two postulates:

$$(i) \quad (x)\neg F(x)$$
$$(ii) \quad (x)\neg(F(x) \to O(x)).$$

T_1 and T' together are logically incompatible. *Their Ramsey sentences are equivalent.* As with Craig's theorem, if the essence of a first-class

---

[7]For a proof of both propositions see Bohnert (1968) and Winnie (1975); an anomaly crops up in case the empirical content of a Ramsey sentence is empty, see English (1973) to whom I am indebted for the example. The Ramsey sentence is examined in many a tract, for instance, in Stegmüller (1973) and Glymour (1980b).

theory is represented by its Ramsey sentence, the underdetermination thesis is false. Incompatibilities in the theoretical superstructure of two otherwise empirically equivalent first-class alternatives appear as artefacts of the theory's formulation.

The Ramsey sentence has been variously interpreted. (1) The "eliminability of theoretical expressions" school holds that since the sentence is stated in observational and logical vocabulary exclusively, "theoretical entities" are eliminated. The Ramsey sentence is one long empirical claim, which captures the essence of a theory. This antirealist view ignores that the second-order dummy variables range over the domain of physical properties; the sentence claims the existence of something or other in the natural world that meets the description. The second view (2) holds that the Ramsey sentence asserts the existence of a model for the theory in question. The Ramsey sentence, as an existential generalization, has realist ontological import according to (2), but the import is less specific than in the original formulation of the theory. Thus Ramseyfication is neutral with respect to the realism issue. Horwich has defended a stronger interpretation (3). He holds that theoretical expressions are "implicitly defined" by the body of the theory. Substitution by a higher-order variable does not change the picture, see section 5.3. Accordingly, the difference between Ramsey sentence and primary theory is that "the latter adds a conventional commitment to call variable properties [...] by the names" used in the full theory (Horwich, 1986, p. 177).[8] The conventionalist interpretation seems wrong. If the difference between Ramsey sentence and its primary theory is just a matter of the choice of *names* then the conjunction $R(T_1) \& R(T\prime)$ should be equivalent to $R(T_1 \& T\prime)$. As the example above shows that this is not generally the case. Incidentally, this lack of transitivity of the Ramseyfication procedure prohibits the explication of the *truth of a theory* as truth of the associated Ramsey sentence Glymour (1980b).

Although the view that the associated Ramsey sentence represents the essence of a theory is compatible with scientific realism (see (2)

---

[8]Compare Horwich (1982, pp. 66–68); Lewis (1984, pp. 78–79) defends a variant account of the Ramsey sentence.

and (3)) and it removes the underdetermination problem, realists will rightly hesitate to take it up. What is lost in Ramseyfication is the specificity of the theoretical claims. It is one thing to claim that electrons have spin 1/2 and another thing to claim that "there exists something that has some property which manifests itself in a such and such way in the presence of a Stern-Gerlach magnet". Realists want to say that theoretical statements are like common sense factual statements in that truth ideally transcends the body of theory in which the sentence is formulated in. This feature is lost in the transition to the Ramsey sentence. Consider the (presumably) theoretical terms "mass" and "electron" in classical electrodynamics and in quantum theory respectively. Both theories can be 'ramseyfied' with respect to these terms. What is lost in the process, is that the electron governed by the laws of electrodynamics *is the very same entity* mentioned governed by the laws of quantum theory. Now, there is nothing that guarantees that the sentences "something moves through the copper wire, that makes [...] the amperemeter register" and "there is something in the helium atom, that compensates for the positive nuclear charge" refer to the same entity: the electron.

2.3 From Denotational Vagueness to Ontological Relativity

The mathematician Adolf Fraenkel was the first to advance "model-theoretic" arguments to demarcate hard limits of our knowledge of the external world. The context is the (then much discussed) problem of the status and role of 'implicit definitions'. The puzzling thing is that implicit definitions do not determine uniquely the meaning of the terms in an axiomatic system. For instance, the term 'line' as a primitive in Hilbert's axiomatization of Euclidean geometry in the plane can be interpreted variously in the naive, visual way or as a line- or circle-segment in the finite "8-point-geometry". After explaining that the truth of a sentence derived in an axiomatic system is invariant under *isomorphic* interpretations and that the axiomatic system by itself

2.3. From Denotational Vagueness to Ontological Relativity

cannot uniquely determine "das inhaltliche Wesen der Grundbegriffe" Fraenkel wrote:

> Diese Auffassung ist auch von größter Bedeutung für den Begriff des "Ding an sich". Wer hinter die uns allein zugängliche Erscheinungswelt das Reich der Dinge an sich setzt, muß diesen die Erscheinungen *eineindeutig und im angeführten Sinn isomorph zugeordnet denken,* so daß wir, ohne die Dinge an sich zu kennen, über sie die für die Erscheinungen gültigen Relationen auszusagen formal berechtigt sind. (Fraenkel, 1928, p. 350, fn. 1; my emphasis)

Fraenkel's point here is that if one postulates a *Reich der Dinge* and their properties-*an-sich* than this system must be an isomorphic copy of the natural model that satisfies an ideal axiomatization of the lawlike relations between the phenomena. Otherwise nothing at all can be said about the *Ding an sich*. Hermann Weyl concurred. He argued that the completeness property of an axiomatic system is to be explicated as isomorphism of its models and went on: "Der Isomorphiegedanke bezeichnet die selbstverständliche unübersteigbare Schranke des Wissens. Auch für die metaphysischen Spekulationen über eine Welt der Dinge an sich hinter den Erscheinungen hat dieser Gedanke aufklärenden Wert. Denn es ist bei einer solchen Hypothese klar, daß die Erscheinungswelt der absoluten isomorph sein muss (wobei freilich die Zuordnung nur in der einen Richtung Ding an sich → Erscheinung eindeutig zu sein braucht) [...]" (Weyl, 1948, p. 22). Substract the speculation about the *Ding an sich*, and what emerges is a plea for structuralism that we have encountered before: it is close to Quine's late views on matters ontological, see section 1.3.

Henryk Mehlberg took this line further. The existence of a range of models which satisfy a given theory demonstrates "the intrinsic denotational vagueness of axiomatic [...] theories presented without appropriate ostensive criteria" (Mehlberg, 1958, pp. 326, 84, 325).[9] Nothing can be said within a theory about the individuals of its universe of discourse:

---

[9] On this point, but in a different context, compare Winnie (1967).

the question as to whether a single predicate does apply to a single individual is never settled by the system, and under these circumstances each of its predicates is completely vague. (Mehlberg, 1958, p. 326)

Mehlberg referred to results by Tarski and Lindenbaum in the famous paper "Über die Beschränktheit der Ausdrucksmittel deduktiver Theorien" (see p. 62). Tarski and Lindenbaum illustrated these facts by application to Hilbert's axiom system of Euclidean geometry: no two points or specific sets other than the empty and universal set can be distinguished by logical and geometrical means, and only a finite number of two-place relations between points can be defined (distinguished) in this axiomatized system. Mehlberg suggested an *operational solution* to the puzzle that denotational vagueness poses. A conditional definition of the theoretical concept like leucocyte is a specification of a preparation procedure (choice of microscope, choice of blood sample, etc.) plus a specification of what an observer is expected to observe in the presence of leucocytes ("a sichle-shaped patch"). There may be other kinds of micro-organisms besides leucocytes that satisfy the 'definition', but a *bound* is established: objects too large, too small, round objects, abstract objects, etc. are all ruled out from the domain of the concept leucocyte. Vagueness remains, but it is reigned in. The argument assumes the common-sense line that there is no 'relativity' in what observation sentences are about, namely "dry, hard goods" (Ryle), at any rate no more than can be expected within the usual vagueness of expressions and ostensive gestures. The next step took John Winnie, who applied a version of the Permutation lemma to the *theoretical vocabulary* of an axiomatized theory in the sciences (see p. 63). He concluded that something is amiss in the classical idea of partially interpreting theoretical terms by way of bridge-laws: theoretical concepts could have non-standard models or referents, while those of the observation sub-language remained constant.[10] Specifically the semantic implications of model-theoretic isomorphisms were further explored by John Wallace in an attempt to refute the view that

---

[10]Note that Putnam's argument against direct realism does not rely on the dichotomy between theoretical and observation language.

reference is physical relationship between terms and extra-linguistic objects: "it is difficult to see how transforming a scheme of reference by permutations of the universe [...] can preserve physical connections" (Wallace, 1979a, p. 314).[11] Finally, there is Putnam's semantical (permutation) argument against naive, direct realism (of which he takes physicalism to be a prime instance), see the next section.

2.4 Semantic Arguments

I review in the present section a class of strong arguments for the *systematic* and necessary underdetermination of theories by the data. They are applicable not only to scientific theories proper, but more generally to *conceptual systems*, provided both are properly formally regimented. The theorems are, the family of Löwenheim–Skolem theorems apart, variants and applications of a result first formulated by A. Lindenbaum and A. Tarski in 1935, but proven in full detail much later. The meta-theorem states that, given the type-theoretic system of logic of *Principia Mathematica*:

> every relation between objects (individuals, classes, relations, etc.) which can be expressed by purely logical means is invariant with respect to every one–one mapping of the 'world' (i.e. the class of all individuals) onto itself and this invariance is logically provable. (Corcoran, 1983, pp. 385–387)

In other words: isomorphic structures cannot be distinguished in the set of sentences of a first order language. The result can be strengthened to the effect, that distinct classes that have equal cardinality and whose complements have equal cardinality are indistinguishable by logical means. This should not come as a surprise. In Euclidean geometry, formulated for example in Hilbert's axiomatization and taking points as individuals, no two points can be distinguished, nor can the direction of

---

[11] Wallace's permutation example and argument is taken up in Davidson (1977, p. 224) and discussed critically in Field (1975) as an instance of "semantic conventionality" that turns out to be philosophically harmless, according to Field.

a line be uniquely determined. The geometric (and logical) relationships between points in the plane, in a triangle say, remain unchanged under reflections and other automorphisms of the whole plane. That is, the statements expressing the various relations between (sets of) points remain their truthvalues under a large class of transformations. In projective geometry *duality* is a familiar fact. On account of the theorem mentioned (which is related to the Isomorphism Lemma), Tarski concludes:

> It is customary to say that our logic is a logic of extensions and not of intentions, since two concepts with different intensions but identical extensions are logically indistinguishable. [...] this assertion can be sharpened: our logic is not even a logic of extensions, but merely a logic of cardinality, since two concepts with different extensions are still logically indistinguishable if only the cardinal numbers of their extensions are equal and the cardinal numbers of their of the complementary concepts are also equal. (Ibid., p. 388)

'Logic' here refers to first- and second order predicate logic (actually to the type-theoretic system of *Principia Mathematica* with certain modifications) but that is unimportant. Tarski and Lindenbaum illustrate these facts by application to the axiom system of Euclidean geometry, noting that no two points or specific sets other than the empty and universal set can be distinguished by logical and geometrical means. Only a finite number of two-place relations between points can be defined (distinguished) in this axiomatized system. For these reasons Mehlberg ascribed to scientific theories, traditionally assumed to be formalizable within first order predicate logic, an intrinsic *denotational vagueness* (see p. 59).

Tarski and Lindenbaum's result proves the strong underdetermination thesis, in a fashion. Consider an empirically successful theory postulating two sorts of non-observational entities e and p, both countably infinite in number. Assume that its first order formulation encompasses a non-observational dyadic predicates ranging over the es and ps of the universe respectively, a predicate applicable to and only to these entities. On the strength of the consideration above it is clear that a

2.4. Semantic Arguments

permutation amongst the *es* and *ps* exists that changes the extension of the predicate in question, *without disturbing the theory's first order observational consequences.* This motivates the following theorem[12]:

Theorem 2.4.1 *Let L a first order language, and T a theory formulated in L, containing the (non-monadic) predicates $P_1, \ldots, P_n$, of which the first $k < n$ make up the observation vocabulary of T. Assume that the extensions of the theoretical and observational predicates do not overlap (S). Let I, the intended model of T (the standard universe), be such that at least one of the predicates $P_{k+1}, \ldots P_n$, say P_{k+1}, is neither empty nor universal. Then there exists a model J of T such that*
(i) I and J agree on the extensions of the observational vocabulary $P_1, \ldots P_k$;
(ii) I and J assign different extensions to at least one predicate $P_{k+1}, \ldots P_n$, i.e. I and J are nonequivalent.

Proof 2.4.1 *Let the (theoretical, l-ary) predicate P_{k+1}^I be neither empty nor universal. Let π be a permutation on the set of "theoretical entities" (D). Define the theoretical predicate P_{k+1}^J such that for $a_1, a_2, \ldots \in D : I \models P_{k+1}^I(a_1, a_2, \ldots)$ iff $J \models P_{k+1}^J(\pi(a_1), \pi(a_2), \ldots)$. This constitutes an isomorphism between the physical interpretations I and J of T. Similarly for the other predicates $k+1, \ldots, n$. From the Isomorphism Lemma it follows that T is true under both interpretations. Select a special permutation π, that maps the extension of P_{k+1}^I onto itself, with the exception of one element: the resulting set is the old one, with one element replaced by its image under π. Hence the extensions of P_{k+1}^I and P_{k+1}^J respectively do not coincide, i.e. the two predicates are nonequivalent.*

The theorem asserts the existence of an *alternative physical model* for the non-observational entities of the theory, one that leaves its empirical content intact. The scope of the theorem is less narrow than the wording suggests. The universe of discourse, the entities in question which are predicated, need not be "thingy" ones. The domain can

---
[12] Adapted from Winnie (1967, p. 226).

be anything, events in space-time (sets of quadruples), or numbers. Fields, potentials and forces, as opposed to countable sets of bodies (lumps of matter) may be construed as predicates of bodies or events. Observational predicates can ascribe immediate, "raw" sensations to regions in the visual field.

The theorem illustrates the slogan that the truth of a set of sentences does not "fix" the reference of its constituent terms. In other words the "truth conditions of whole sentences undermine reference". This prototype of semantic proofs of the strong thesis of observational indistinguishability is due to John A. Winnie, who argued against a then popular theory of how theoretical terms get their cognitive meaning: by being *partially interpreted* through the observational consequences of the theory and its correspondence rules. Non-observational concepts were thought (e.g. by Carnap) to derive their cognitive significance solely from the way they are embedded in the surrounding body of theory. The next theorem sharpens the previous result in that there is a *numerical* interpretation of T, on the domain of natural numbers.

Theorem 2.4.2 (Phytagoreisation) *Let L be a first order language, and T a theory formulated in L, containing the (non-monadic) predicates $P_1, \ldots, P_n$, of which the first $k < n$ make up the observation vocabulary of T. Assume that the extensions of the theoretical and observational predicates do not overlap (S). Let I, the intended model of T (the standard universe), be such that at least one of the predicates $P_{k+1}, \ldots P_n$, say P_{k+1}, is neither empty nor universal. Then there exists a model J of T such that*
(i) I and J agree on the extensions of the observational vocabulary $P_1, \ldots P_k$;
(ii) J assigns a set of arithmetical entities as extension of the predicates $P^J_{k+1}, \ldots P^J_n$.

It might seem that the argument is simply deficient: non-observational entities are located in space-time, numbers are not. Hence an appropriate constraint added to the theory should eliminate at least the dreary possibility of Pythagoreization. Such a constraint is implemented by adding to the body of the theory suitable axioms of geometry if they are not already in place, which the entities are supposed to satisfy.

However, this move would amount to adding more theory, and hence will not change the conclusion (Winnie, 1967, p. 229).[13] Difficulties arise from the assumption (S) in the theorem just stated that proper observational predicates apply to observable entities only, and theoretical predicates apply to non-observational entities only. S allows to permute the sub-domain of non-observable entities and reinterpret the theoretical predicates exclusively, i.e. without disturbing the observational substructures. It practically *defines* non-observational entities as those objects, of which no observation predicate is true of. However, S seems to be factually wrong as David Lewis pointed out. A H_2O molecule is a theoretical entity (on any standard account of theoreticity). Yet, in sufficient large amounts it becomes an observable entity in bottles and ponds, of which such homely predicates like "is cold" can be said to be true of.

Redrawing the extensions of some of the theory's predicates changes their "meaning". This consequence is, perhaps, acceptable for theoretical predicates proper, but what about mixed predicates, predicates denoting entities from the extensions of the $P_1, \ldots, P_k$, as well as of the remaining predicates? If the predicate takes values on non-observable entities it is, or has become, a theoretical predicate. Its meaning will change under a suitable chosen one-one mapping of the domain for non-standard models, but that may be considered acceptable since one is dealing with a theoretical predicate. Perhaps pure *extensionality* is the crucial assumption, the neglect of intensions that lead to the counter-intuitive result? We will see below that the theorem holds in the framework of Carnapian semantics.

The main difference between Winnie's theorem and the next theorem – different objectives namely standard accounts of meaning and realism respectively apart – is that intensions of predicates are taken into account. The collection $(I^r), r = 1, 2, \ldots$, assigns an intension to every predicate of L, in the sense of Carnap's semantics. Formally, Putnam's theorem is a straightforward generalization of the above.

Theorem 2.4.3 (Putnam) *Let L be a first-order language, containing the predicates (not necessarily monadic, but not mixed) $P_1, \ldots, P_n$,*

---

[13]Putnam objects in the same way to the call for causality in fixing reference.

of which the first k make up the observation vocabulary of L. Let I^r, the intended model of L (the standard universe) in world number r ($r = 1, 2, \ldots$), be such that at least one of the predicates $P^r_{k+1}, \ldots P^r_n$; is neither empty nor universal. Then there exists an interpretation J^r of L, such that, for all worlds r ($= 1, 2, \ldots$) :
(i) I^r and J^r agree on the extensions of the observational vocabulary $P^r_1, \ldots P^r_k$;
(ii) I^r and J^r assign different extensions of at least one predicate $P^r_{k+1}, \ldots P^r_n$; i.e. they are nonequivalent.

Proof 2.4.2 *Repeat the steps of theorem 2.4.1 above for each world r.*[14]

Note, that nothing essential depends on the first-order characteristic of the language L here; analogues hold for higher order languages.

As to the interpretation of the lemma: it illustrates – like Tarski's theorem did – that truth conditions of whole sentences undermine reference (Putnam, 1981, p. 35). According to Putnam we go wrong in equating understanding a language with knowing the truth conditions for sentences. The result reflects two of Quine's earlier theses, the "indeterminacy of reference" and "ontological relativity" thesis, examined earlier. Note also, Putnam's lemma does not dispense with the difficulties related to assumption S from a few paragraphs before (see the second of three remarks in Putnam, 1981, p. 218).

This series of theorems culminates naturally in the Skolem and Löwenheim theorem. As is well known, the "downward" version asserts that for any theory in *first order* language a number theoretical interpretation exists. If a first-order theory has a non-denumerable model M at all, than (i) it has a countable model whose *domain* is a subdomain of M, and (ii) the *relations* of the countable model are obtained by restricting the relations of M to the domain of the countable model. In other words, truth is preserved even though the universes of discourse underlying the theory in question can be radically different.

---

[14]See Putnam (1981, pp. 26, 217); compare Putnam (1977, pp. 12, 15–16) and Putnam (1983b, pp. 12–14).

2.4. Semantic Arguments

The theorem in question is discussed in every standard logic text and many papers. I provide the statement for completeness' sake:

Theorem 2.4.4 (Skolem, Löwenheim) *If a set of formulas of first-order predicate logic has any model at all, then it has a model with a denumerable infinite domain.*

If a first-order theory has a *non-denumerable* model it cannot be *categorical*, i.e. not all of its models can be isomorphic. Consistent theories formulated in first-order logic have models with domains of *different cardinality*. In view of the relation between models and the ontology of a theory it is worth noting, that the plethora of models a consistent first-order theory admits is not an unstructured or "anarchical" set. Models in this set may contain smaller submodels, and there are *minimal models*.[15] The Löwenheim–Skolem theorem's "upward" version says that any consistent set of formulas of first-order predicate logic has models of every infinite cardinality.

The theorem is often taken to imply that no first-order formalization of set theory can capture the notion of "set", since as a consistent theory it admits non-standard models with denumerable domains. Thus the axioms 'underdetermine' the notion of set in a sense. However, given the *usual understanding* of the $\in$ relation, the Zermelo–Fraenkel system of set-theory does capture our concept of set completely. In other words, concepts like "$\in$", "finite", etc., have only relative meaning. The theorem's general import has been much discussed and this is not the place to review the discussion.[16]

The theorem does not apply to second order languages, often considered necessary for the formalization of all of higher mathematics and

---

[15] One can make statements regarding the number of elements in the domain of any model of the theory within first order predicate logic *with* the identity relation, see Hilbert and Ackermann (1959, p. 104) (§9). The completeness theorem in case guarantees the existence of a model which is *either finite* or denumerable. There are expressions in the first order predicate calculus that can be satisfied in every finite domain, but not in the natural numbers under the usual interpretation of relations among them, ibid. p. 123 (§11).

[16] See Myhill (1951) and Hallett (1994) for an exposition with respect to Putnam's claims about realism; and p. 158 for some remarks on 'finiteness'.

hence for the physical sciences. Partly because of objections stemming from this fact Putnam relies for much of his arguments against metaphysical realism on the weaker Permutation lemma.[17] A generalization of the theorem goes like this: in a second order logic with identity, if a set of formulas has any models it has models whose domains are all either finite or denumerably infinite. A formulation in *second-order logic* renders many important theories *categorical*: their principal models are alike in structure. This feature is absent in first-order theories without identity. The *primacy of standard first-order logic*, argued Quine, rests on its having a consistent and effectively defined set of axioms from which all of the valid formulas are derivable or "a complete proof procedure" (Quine, 1969a, p. 111). He conceded that *alternative logical schemes*, whether or not they share this characteristic, can import *alternative* "concepts of existence" (ibid. p. 113).

It would be mildly embarrassing if weighty metaphysical claims, as well as the truth of U, hinge on a choice between formalization in first- or second-order logic. I discuss the implications and significance of the logico-mathematical facts reviewed here for the truth of the underdetermination thesis, in particular U, in section 4.2.

2.5 Physical Equivalence

This section looks at ways in which equivalencies arise and are formulated in mathematical physics.

The existence of equivalence relations (between 1-forms, between Lagrangians, between sets of differential equations under co-ordinate transformations, etc.) is closely connected to the existence of invariances and symmetries, and widely investigated. "Symmetries" of physical theories are either *external* symmetries of a system, involving transformations of the space-time coordinates (Galilean transformations; rotations; conformal mappings; reflections, etc.), or *internal* symmetries, which transform the objects of the theory without changing their co-ordinates (gauge transformations; supersymmetric trans-

---

[17] Putnam comments on the role of second-order logic in Putnam (1989).

2.5. Physical Equivalence 69

formations, etc.). I will focus here on classical mechanics in its Lagrange formulation. The advantage is that the Lagrange formalism has application in other areas of physics, as the Lagrangian formulation has been transferred from mechanics proper to classical and quantum field theories, relativistic and non-relativistic alike. The question of "physical equivalence" presents itself in three ways: (1) the equivalence or non-equivalence of different mechanical models within the Lagrangian framework for a given set of phenomena; (2) the equivalence or non-equivalence of the three classical formulations of mechanics: Newtonian mechanics, the Lagrange formalism, and the Hamilton and Hamilton–Jacobi formalisms; (3) the equivalence, empirical and theoretical, of different axiomatizations of classical mechanics. From a study of the answers to question (1) I try to catch in an exemplary way the idea of "physical equivalence" between theories.

A basic notion for our purposes is that of the trajectory in the configuration space of the mechanical system. The configuration space is the "state space" of the system, the set of possible dynamical states. Knowledge of the trajectory, given explicitly and parametrized by the time parameter, allows to compute all other observables of the system. The system's trajectory (a time-evolution) should not be confused with the "graph" (a path) of the trajectory in configuration space. In mechanics the system's trajectory in an n dimensional configuration space is given as the unique solution of a set of n second-order differential equations with initial conditions, the Euler–Lagrange equations. Suppose, the trajectories of two mechanical systems (in a given chart, if we are dealing with a manifold) are identical, starting from identical initial data. It is natural to stipulate that the two systems are "physically equivalent" under these conditions, particularly when this holds for all initial data. The stipulation generalizes and liberates the stricter positivistic definition of empirical (deductive) content, but is not a criterion of full or theoretical equivalence.

The notion of "physically equivalence" will be made precise in order to state some propositions on equivalent mechanical models. The presentation will have a qualitative character. I suppose the manifold M is given together with a (collection of) local charts ϕ (along with a metric in order to form invariants, but it plays no role in the following). Generalized coordinates in configuration space are denoted by q_i, $i = 1, \ldots, n$ (or collectively as $\mathbf{q}$), where n is the degree of freedom of the system under investigation. However, I am not interested in the most general case, so the objects are mechanical systems in Cartesian space without (geometric etc.) constraints. The underlying manifold is then simply R^n, and for a set of N interacting particles say, $n = 3N$. A trajectory may be defined as an ordered pair of the underlying manifold M and a mapping $U_t : R \to M$, the time-evolution or trajectory in M, where the dependence of U_t on initial data $(\mathbf{q}, \dot{\mathbf{q}})$ in a given local chart ϕ is suppressed: $\langle M, U_t \rangle$. The trajectory is parametrized with time, and is supposed to be extendable from any interval in R to the the whole of R, for simplicity's sake. Given two trajectories in local charts which differ only by a *global* co-ordinate transformation between the charts on M cannot be distinguished by physical means. They represent the same underlying trajectory on M. This intuition can be captured by stipulating that two trajectories $\langle M, U_t \rangle$ in ϕ and $\langle M, U_t \rangle$ in ϕ are "equivalent" if and only if there is a differentiable bijection between ϕ and ϕ'. Note, the fact that trajectories are those of classical mechanical systems has played no part in the previous definitions. (Strictly speaking, the system's state space is not the manifold M, but the space of smooth mappings $\gamma : R \to M$.)

Knowledge of the mechanical system's Lagrange function and the trajectory derived from it characterizes the system completely. Of particular interest is the celestial mechanics paradigm, i.e. of a system of point particles interacting through conservative pair-potentials. Hence, the following stipulation:

Definition 2.5.1 *A N-particle "Newtonian" Lagrangian mechanical model $\langle M, U_t, \mathcal{L} \rangle$ is a (smooth, connected) differentiable manifold M together with a Lagrange function $\mathcal{L}$ defined on its tangent bundle TM such that:*

1. $M = R^{3N}$ in ϕ.

2. $L : R^{3N} \times R^{3N} \times R \to R$, with $(\mathbf{q}, \dot{\mathbf{q}}, t) \mapsto \mathcal{L}(\mathbf{q}, \dot{\mathbf{q}}, t) = T - V$.

3. $T : R^{3N} \times R^{3N} \to R$, $T = T(\mathbf{q}, \dot{\mathbf{q}})$ *is homogeneous and of second degree in $\dot{\mathbf{q}}$.*

2.5. Physical Equivalence

4. V is a function of **q** *alone.*

5. The Euler–Lagrange equations hold.

T denotes the kinetic energy of the system of n point-particle, and V the conservative interaction potential. The sum of both quantities can be shown to be conserved over time and equals the system's total energy. The definition can be modified and generalized in obvious ways, to take account of non-Euclidean manifolds, constraints, and Lagrange functions which do not split in potential and kinetic energy.

In accordance with the considerations above two such models are physically equivalent if they are solution-equivalent. That is, the trajectories in configuration space derived from the respective Euler-Lagrange equations are identical for all initial values, modulo global co-ordinate transformations. The property of a trajectory of "being a dynamical solution" of the system's Lagrange function is what is preserved in this case, analogous to the order-preserving isomorphism between formal structures (see above). I do not offer this explication of equivalence between mechanical models as the final word on equivalent theories in physics; for that the scope of the proposal is too limited. Yet, it captures the intuitions of scientists, I believe, and complements the notions of interdefinability discussed earlier:

Definition 2.5.2 *Two mechanical models $\langle M, U_t, \mathcal{L} \rangle$ in a local chart ϕ and $\langle M, U_t', \mathcal{L}' \rangle$ in ϕ are physically equivalent if and only if the trajectories $\langle M, U_t \rangle$ and $\langle M, U_t' \rangle$ are equivalent in the sense defined above.*

Consider the two-body problem in a potential that depends only on the relative distance between the point masses. It can be rewritten as a *one-body* problem for a particle with a "reduced" mass in the same potential. It follows from our definition of physical equivalence that the two Newtonian models are *not* equivalent since there is obviously no diffeomorphism between $M_1 = R^6$ and $M_2 = R^3$. On the other hand the one-body model is just a technical convenience with just the same set of dynamical solutions. However, the models *are* nonequivalent if looked at independently, and hence the proposed definition is in accordance with intuition.

These clarifications and preparations ease the statement of a number of facts about equivalencies between mechanical systems. First, consider two systems of charts on M, with co-ordinates $\mathbf{q}$ and $\mathbf{q}'$, and two Lagrange functions $\mathcal{L}(\mathbf{q}, \dot{\mathbf{q}}, t)$ and $\mathcal{L}'(\mathbf{q}', \dot{\mathbf{q}}', t)$. One wants a sufficient condition for when two such mechanical systems are really two systems, instead of one system in different local charts. The following proposition gives the answer:

Theorem 2.5.1 *Two Newtonian mechanical systems $\langle M, U_t, \mathcal{L} \rangle$ and $\langle M, U'_t, \mathcal{L}' \rangle$ are physically equivalent if*

1. *The coordinate transformation $\psi = \phi' \circ \phi^{-1}$ is a diffeomorphism.*

2. $\mathcal{L}' = \mathcal{L} \circ \psi$.

In other words, if $\mathbf{q}(t)$ or rather $U_{t,\phi}$ is a "solution" of the Euler–Lagrange equations associated with $\mathcal{L}$ describing the system's time-evolution, then $\psi(\mathbf{q})$ or $U_{t,\psi \circ \phi}$ is a solution of $\mathcal{L}'$ in the new coordinates, and vice versa. The proof is omitted; it turns on a straightforward rewriting of the various derivatives in $\mathcal{L}$ in the new coordinates using Leibniz's rule. Because ψ is one-one and both it and its inverse are differentiable, the "transformation matrix" between the variational derivatives $(\partial \mathbf{q}/\partial \mathbf{q}')$ is not singular.

Consider these two one-particle Lagrange densities:

$$\mathcal{L}_1 = \dot{q}^2 + q^2$$

$$\mathcal{L}_2 = (\dot{q} + q)^2$$

Each of the two determines by way of a classical action principle and the Euler–Lagrange equations the motion of a point particle of mass 1 in an external field. $\mathcal{L}_2$ differs from $\mathcal{L}_1$ by positing an additional, velocity-dependent "potential term". Yet, both lead to exactly the same particle trajectories, subject to the same boundary conditions, as can be easily verified by doing the Euler–Lagrange derivatives directly. The deeper reason is that the difference $\mathcal{L}_1 - \mathcal{L}_2$ is a total time derivative of

2.5. Physical Equivalence

a function depending only on the co-ordinate. Lagrange functions (and Lagrange densities in the case of fields) differing in a total differential in time give rise to the same observational dynamics.

Theorem 2.5.2 *Two Newtonian mechanical systems $\langle M, U_t, \mathcal{L} \rangle$ and $\langle M, U'_t, \mathcal{L}' \rangle$ in a given chart are physically equivalent if*

$$\mathcal{L}' = \mathcal{L} + \frac{d}{dt}(\eta).$$

The mechanical Lagrange function turns out *not to be unique*, different functions give rise through the Euler–Lagrange equations of the same set of dynamical trajectories. The Lagrange function is a *gauge quantity*. Consider for instance the classical motion of a charged particle in electric and magnetic fields. As is well known these fields can be expressed in terms of a scalar and vector potential Φ and $\mathbf{A}$, resulting in the second-order Maxwell equations. Let η be a scalar, twice differentiable function of position and time. Then the gauge transformations of the potentials

$$\Phi_\eta(x, y, z; t) = \Phi(x, y, z; t) - \frac{1}{c}\dot\eta(x, y, z; t)$$

$$\mathbf{A}_\eta(x, y, z; t) = \mathbf{A}(x, y, z; t) + \nabla \eta(x, y, z; t)$$

induce a gauge transformation of the corresponding Lagrangians:

$$\mathcal{L}_\eta(x, y, z; t) = \mathcal{L}(x, y, z; t) + \frac{d}{dt}\left(\frac{e}{c}\eta\right)$$

Both Lagrange functions return the same particle dynamics. They are properly equivalent. The potentials, as against the uniquely determined field-strength, cannot represent objective features of the world. Similar results regarding the non-uniqueness of the Lagrangian hold for the standard Lagrange formulation of any (relativistic) field theory.

If the condition of (strict) convexity in the Lagrange function with respect to the generalized velocities is violated, a Lagrange function may give rise to corresponding Euler–Lagrange equations that are *inconsistent*, e.g. $\mathcal{L} = \dot{q} + q$. Conversely, there are classical physical systems for which no Lagrangian formulation exists to derive the dynamics from (e.g. the second-order Proca-equations for a relativistic massive spin 1 "Proca" field do not have a natural second-order Lagrangian density). Note, the uniqueness of the system's trajectory is guaranteed by well-known theorems on initial value problems; however, the boundary value problem for the dynamics of a mechanical system (fixing the position for two different times), in some respects a more natural formulation of mechanical problems, will not in general permit a unique solution. The one-particle Lagrangian $\mathcal{L} = \dot{q}^2 - q^3$, with $q(t_1)$ and $q(t_2)$ ($t_1 \neq t_2$) given, has a countable infinite set of solutions. (The difference between the two settings is not adequately mirrored in our definition of a mechanical model.)

Somewhat between internal and external symmetries and the corresponding p-equivalencies of mechanical systems lies the *kinematical equivalence* of dynamical solutions. Two trajectories of the same system will appear different for observers in different states of motion. If the apparent differences between two trajectories can be explained in full by taking into account the different states of motion of two observers, the solutions are called kinematical equivalent. The formal realization is given in mechanics by demanding invariance under the Galilei- or Lorentz group. Since this type of equivalence leads to a trivial (observer-relativized) notion of underdetermination and is examined copiously in the relativity literature I will not discuss kinematical equivalence any further.

The *third equivalence property* of classical mechanics is the "equivalence" of the Lagrangian and the Hamiltonian formulation of classical particle mechanics. As above, the proposed notion of a mechanical model will center on the set of "trajectories" associated through Hamiltonian equations instead of Lagrange equations with a given mechanical system. While the $\dot{\mathbf{q}}$ are tangential vectors ($\in TM$) to the configuration manifold M, the generalized momenta, defined as $\mathbf{p} = \frac{\partial \mathcal{L}}{\partial \dot{\mathbf{q}}}$, are cotangent vectors, elements of the cotangent bundle T^*M. T^*M of dimension $2n$ has a symplectic structure, which in suitable local coordinates, is given by the non-degenerate, closed, skew-symmetric 2-form $\omega = \sum_i dp_i \wedge dq_i$. This structure is natural (= unavoidable) in the sense, that for any closed 2-form on T^*M there is a local chart in which the 2-form can be represented in the way stated (Darboux's theorem). Apart from ω, the second defining characteristic of a mechanical model of course is the Hamiltonian H, a smooth function from the cotangent bundle into the reals. The exterior derivative dH is a differential 1-form, which in local coordinates is equal to the total differential of H.

A *Hamiltonian* vector field ζ_H is defined by the condition:

$$dH(\eta) = \omega(\zeta_H, \eta)$$

for all vectors η tangent to T^*M_x at a chosen point x. The mechanical system's trajectories – the "phase flow" – can now be shown to be the *integral curves* of the vector field ζ_H. These are the dynamical "solutions" which are critical in determining the p–equivalence of two mechanical models. Hence the following definition of a Hamiltonian mechanical model in geometric terms:

Definition 2.5.3 *A "Hamiltonian" mechanical model $\langle M, \omega, U_t, \zeta_H \rangle$ in a local chart ϕ is a (smooth, connected) differentiable manifold M together with a 2-form ω on the tangent bundle such that:*

1. *ω is non-degenerate, closed and skew-symmetric.*

2. *U_t is the time-evolution of the integral curves associated with the Hamiltonian vectorfield ζ_H.*

The Lagrangian–Hamiltonian equivalence is treated in textbooks as a straightforward matter. Assuming the Lagrange function is a positive definite quadratic form with respect to the velocities (or a convex function), the mechanical system's Hamiltonian is the *Legendre transform* of the system's Lagrangian viewed as a function of the particle velocities. The Legendre transform maps the Lagrange function on the tangent bundle to the Hamilton function on the co-tangent bundle; it does not depend on the chosen coordinate system. Though it is relatively easy to verify the equivalence of the two systems of differential equation, a rigorous proof of the equivalence appears somewhat elusive. Batlle et al. prove the equivalence for the case of *constrained* mechanical systems, for instance, systems like the simple pendulum, particles on a circle or on any other constraining surface, etc. C. Batlle et. al (1986). No general rigorous proof appears do exist yet for the unconstrained case.[18]

The matter of observational equivalence is a subtle one. In the Hamiltonian formulation a particle's position and momentum are treated as conceptually independent dynamical quantities. (The independence becomes particularly clear in the standard mathematical formulation to quantum mechanics.) It is only through Hamilton's equations that the time-derivative of the particle position becomes "correlated" with the particle momentum. The relation between the two is *law-like*, an *empirical fact* of the mechanical world. In the usual Lagrange formulation, however, the identity $p_{x,i} = \frac{\partial H}{\partial \dot{x}_i}$ (the "generalized momentum") is a *definition*. The definition needs to be *added to* the Lagrange formulation, and turns the momentum into a derived, secondary quantity. Note, while the Lagrange function is not unique, a mechanical system's Hamiltonian is unique.

I can only touch on the question of how *Newtonian classical mechanics* and Lagrangian mechanics are logically related. Despite textbook treatments suggesting perfect empirical equivalence, the matter is complicated through the fact that the basic entities posited in the two mechanics, forces versus energy and Lagrange function, particle positions versus degrees of freedom, are of different conceptual character. Indeed, the two turn out not to be "equivalent" in the Sneedian structuralist account of (theoretical) equivalence and empirical equivalence, at least outside the family of conservative pair-potential models, particular celestial mechanics (Balzer et al., 1987, pp. 294, 303). It is straightforward to show that a given model of Newtonian mechanics is matched by a Lagrangian model, but the converse can be difficult to establish in general, making the situation look like a one-sided reduction. However, I suspect that the negative result is rooted not in the physics, but partially in the particulars of the Sneedian account. The prospects of proving in the structuralist framework more elaborate

---

[18]See C. Batlle et. al (1986, p. 2953) and references therein; my presentation follows closely Scheck (1990), who discusses the (likely) equivalence in formal terms in §5.6.

examples of equivalence, like the equivalence between classical and the covariant (tensor) formulations of electrodynamics, are equally dim. The "semantic account" sketched above, on the other hand, appears readily transferable from mechanics to quantum mechanics and other theories, and is very intuitive. One problem, however, is that the "ontology" of a mechanical system is only poorly reflected in the notion of a model. A commitment to N point-particles, their masses and positions will be acceptable to scientists, but a commitment to the whole $3N$ dimensional state space or the Lagrange function is less acceptable.

I tentatively identified above the mechanical system's empirical content with the system's trajectories, from knowledge of which all other quantities can be calculated (ideally). Sometimes this choice of what counts as observable is implausible, for instance in case the system of n masspoints represents a system of (non-observable) elementary particles. Suppose $Q_1, \ldots, Q_l$ is a set of independent (complete) time-dependent observable (macro) quantities, whose empirical (causal) interrelations are known in a time interval I. Let us call this system a *data-model* for short. Suppose further, that these interrelations are expressible in form of a system of equations, a system of well-behaved second-order differential equations ($\ddot{Q}_i = f_i(Q_1, \ldots, Q_l; \dot{Q}_1, \ldots, \dot{Q}_l; t); t \in I; i = 1, \ldots, l$) in particular. (This assumption is in many cases an unrealistic one.) If the observables can be expressed in terms of solutions of a microscopic mechanical model $\langle M, L \rangle$, such that the resulting set of equations is consistent with the Euler–Lagrange equations, the mechanical model may be said to "explain" the data model. In order to avoid quarrels about the proper notion of explanation, let us say that in this case $\langle M, L \rangle$ is empirically adequate. However, the n-particle microscopic system will in general not be unique for bounded time intervals I. Naturally the question arises, (1) under what conditions a given (complete) data-model can be accounted for by some Newtonian Lagrangian model $\langle M, L \rangle$; and (2) provided there is such a model, whether it is in a sense *uniquely determined* by the data model and some general assumptions about the mechanical model. Otherwise the data-model underdetermines its micro-mechanical explanation. There are no general answers to these questions, but Poincaré (1952, p. 221) suggests that the answer to the second question is negative and that this can be proven.[19] Poincaré argued that his "lemma" explained the historical de facto existence of (weakly) empirically equivalent mechanical explanations of electrostatic phenomena in the literature of his time. He suggested that the lemma on the non-uniqueness of mechanical explanations showed that the then ongoing search for an *experimentum crucis* between the "two fluid" and "one fluid" account of electrostatics was in vain. Important or informative is the general structure of a mechanical model, which frames the phenomena, not what concrete models look like.

There is indeed a large, distinct class of cases where different mechanical models give rise to the same trajectories and so are fully p-equivalent: those which are due to global symmetries of the theory, like permutation, rotation, translation, time-reflection symmetries, or to "local" gauge symmetries.[20] Individual models of a given phenomenological process may differ on statements regarding the value of parameters or states which are subject to a symmetry property and so are strictly incompatible. The *theory itself* tells the scientist that no amount of deductive data can single out one model from among a set of alternative models – relative to the theory in question. It is tempting – and to a degree established scientific practice – to ontologically "discount" symmetry related differences between models. However, the temptation should be resisted. Newtonian models of planetary motions, say, may attribute different absolute motions of the center of the universe, but are nevertheless observationally equivalent. The reason is the underlying Galilean symmetry. The models are physically (p-)equivalent in our sense, but few scientists, historically speaking, felt inclined to reject the quantity 'velocity of the center of mass of the universe' as a "physically" meaningless one. Moreover, through the *conjunction* of Newtonian mechanics and classical electrodynamics the quantity in question can be determined by observation (van Fraassen, 1980a, p. 49). The conjunction "option" is always present: the possibility that the symmetry may be broken (and the parameter

---

[19]See Poincaré's introduction to his *Electricité et Optique* (Poincaré, 1901, pp. vi f.).

[20]Note: most global symmetries do not actually hold or hold only approximately.

in question be actually measured or inferred) in a more comprehensive theory, which may be simply a conjunction of the original theory with another, perhaps novel theory.

Waiving the conjunction option for the moment, if all cases of strong underdetermination between alternatives had their origin in symmetry properties of a theory, the skeptical problems raised in chapter 1 could be resolved easily. Unfortunately, main examples of underdetermination (in physics), of the strong or transient kind, cannot be reduced to instances of an internal symmetry or gauge freedom. There are, for example, *axiomatic approaches* to classical particle mechanics that differ prima facie in the presupposed ontology, yet have the same global dynamical symmetry groups. In J. C. C. McKinsey and Suppes (1953) the particles' masses, locations and forces acting between them are treated as mutually independent (irreducible) concepts, while in Mackey (1963, pp. 2f) and SantAnna (1996) forces are defined (reduced) and derived from directly observable quantities in a way that would be recognizable to Ernst Mach or Heinrich Herz. Not all manners of 'defining a quantity away' should count as genuine ontological reductions, as Wilson (1981) illustrates, yet the line between those that do and those that do not is drawn case by case with varying plausibility (see below). One criterion for choosing a set of axioms for classical point mechanics from among alternatives would be conformity or closeness to scientific practice, which may favour the McKinsey–Suppes system of axioms. But neither is this conclusion foregone, nor is it particularly evident why 'conformity' to changing and changeable patterns of practice should count as a sure guide to matters ontological.

Turning from mechanics to electrodynamics, the Wheeler–Feynman theory of electrodynamics, presented in Wheeler and Feynman (1949), appears to be a genuine, observationally equivalent alternative to standard electrodynamics with Lorentz's force. Putnam in Putnam (1978) cites this action-at-a-distance theory as an instance of the (strong) underdetermination of a mature, well-tested theory. Classical electrodynamics postulates, besides charged particles, two sorts of physical vector fields spread out in space, governed by Maxwell's equations, which act directly on (point-like) charged particles and vice versa.[21] The alternative explains electromagnetic phenomena with the help of so-called (in time) "retarded" and "advanced" Liénard–Wiechert solutions to Maxwell's equations and action-at-a-distance interactions (forces) between charged particles, derived from the former. Absent are the familiar electromagnetic vector fields as independent, primitive physical objects; hence the "ontology" of the Wheeler–Feynman theory appears to differ from that of the standard theory, and the difference is unrelated to symmetry properties of the theory. This radical reconstruction of the classical theory rests on taking causally anomalous "advanced" solutions into account: *future* values of physical magnitudes are allowed to influence the present state of a charged particle. Feynman and Wheeler thus resolved a conceptual difficulty of the classical theory, the self-energy problem: the action of the charged point-particle on itself prohibits that the energy of its electro-magnetic field is finite. Among their assumptions is that an accelerated charge radiates off energy only if in the presence of absorbing charges located somewhere in space. The retarded potential of the primary charge accelerates the absorbers, which in turn radiate half of their advanced potential "backwards in time" to the primary charge, the source, just when it began to accelerate. The net effect is a force on the original charge, just at the time when its acceleration began, which produces a damping effect that keeps the self-energy finite.

A number of remarks are in order. *First*, since classical electrodynamics gives rise to a formally divergent series, it is mathematically *inconsistent*. Wilson observes correctly that this fact *rules out intertranslatability*, as criterion of identity between alternatives, with the (presumably consistent) Wheeler–Feynman theory (Wilson, 1981, p. 421). So the alternatives are genuine, and strongly observationally underdetermined, if only the deductive content is taken into account (the conceptual differences are not symmetry induced). *Second*, the Wheeler–Feynman theory does not seem to be the ontologically more

---

[21] In a fully relativistic formulation a single four-by-four tensorfield replaces the magnetic and electric field vector. To simplify things, only the vacuum case is considered here, not interactions of the fields and particles in and with a material medium.

parsimonious of the alternatives, since it assumes besides charged particles forces between those charged particles, provided the Liénard–Wiechert solutions of Maxwell's equations themselves are taken as mere mathematical auxiliaries.[22] *Third*, in Maxwell's electro-dynamics the fields themselves carry energy and momentum, and the corresponding conservation laws for energy and momentum are satisfied for the total system of particles and fields. The same does not hold for the Liénard–Wiechert potentials. It is tempting to turn the niece properties of the fields in Maxwell's theory into an argument for their 'reality' and against the truth or likelihood of the alternative. But let us keep in mind that an electromagnetic field is an unusual kind of object in that it has an infinite number of degrees of freedom. A theorem by H. Van Dam and E. Wigner states that, under mild restrictions, in theories, which postulate action-at-a-distance interactions of particles the total amount of energy and momentum of such a system of particles is not constant in all inertial frames (Dam and Wigner, 1996). Nevertheless, Lorentz invariant instantaneous action-at-a-distance theories have been constructed, which turn out to have necessarily some unusual features (like many-body forces) (see Hill, 1967). Hill also raises the question whether there are action-at-a-distance theories for which no field theoretic alternative exists. *Fourth*, although an inconsistency is an excellent reason to reject a theory, particularly if an observationally equivalent alternative is at hand, the scientific community has in the main stuck with the orthodox theory. An explanation may be found in the introduction of counter-intuitive, causally anomalous "advanced" solutions of Maxwell's equations; of an unabashedly action-at-a-distance mechanism[23]; and perhaps in the "localized" way the original problem presents itself, namely in form of a divergent series whose first terms are perfectly respectable, which does not impede directly the theory's applicability. It is controversial in how far the first two reasons carry any evidential weight, since the 'a-causal'

---

[22]Properly so called ontological questions play a shadowy role in scientific progress. For instance, for the revolutionary transition from classical mechanics to special relativity theory the material ontology was irrelevant Gaifman (1975).

[23]See Mundy (1989) for an account of action-at-a-distance versions of classical electro-dynamics and their historical reception.

processes cannot be observed in principle and presumably both 'versions' of electro-dynamics make exactly the same predictions.

2.6 Underdetermination of Geometry

Consider Weyl and Reichenbach's contention that the geometry of space and time and the laws of particle dynamics are tested together as a unit. Denote summarily by F the physical theory in question, and by G the underlying *orthodox* geometric assumptions. A "universal" force is defined as a force that acts on all materials in the same way, and no object can be shielded or isolated from its influence. (Different from Newtonian gravitation, universal forces need not have matter as a source.) Physical space can be said to have a specific geometry only relative to the choice of a standard of length (relation of congruence); space does not have a natural intrinsic metric, according to Reichenbach. Admitting a "universal" force to F which affects (deforms) the measuring instruments, measurements of distances, lengths, areas and angles will result in a different geometry, G' say. Specifying either directly the standard of length, or the form of universal force (and hence the rigidity of measuring rods) is, if we follow Reichenbach, a coordinative definition, i.e. a conventional decision. Through mutually compensating adjustments the resulting cluster of laws $F'+G'$ becomes indistinguishable from the orthodox $F + G$ with regard to deductive empirical content, basic observations like particle positions and times. (To illustrate the claim that physical laws, the equation of motion for instance, have no "absolute" or intrinsic meaning, consider the curve in the plane corresponding to the equation $y = \sin x$. The equation describes indeed a *sinus curve* in the standard metric $ds^2 = dx^2 + dy^2$ (ds is the infinitesimal line element.) The equation however corresponds to a *circle* if the metric of the plane is $ds^2 = y^2 dx^2 + dy^2$.) The introduction of "universal" forces, so unlike most forces we know, may not be necessary to achieve the same purpose. Putnam argues that alternative geometry+force pairs can be constructed *without "universal" forces* if only an appropriate "interaction" between differential forces

2.6. Underdetermination of Geometry

is postulated Putnam (1975b).

Stated in a more formal way this becomes clearer. A geodesic in a curved space describes the inertial, free fall trajectory of a particle; it is a generalization of the "straight line" in the Euclidean space, with the characteristic property of being the shortest between two points. The motion of a test particle moving along a geodesic is given by the equation

$$\frac{d^2 x^a}{dt^2} + \Gamma^a_{bc} \frac{dx^b}{dt} \frac{dx^c}{dt} = 0, \tag{2.1}$$

where the Christoffel symbols of the second kind Γ^a_{bc} ($a, b, c \in \{1, 2, 3, 4\}$) are functions of the metric on our space and its partial derivatives (or components of a Riemannian connection). Suppose we have found a set $\{\hat{\Gamma}^a_{bc}\}$ such that the observed particle motions satisfy that equation; hence have "discovered" the geometry of space. (For instance, by observing the inertial, straight-line motion of billiard balls between impacts one would easily be led to postulate that all components of Γ^a_{bc} vanish everywhere, and that the billiard is a plane billiard.) Suppose that researchers are led to believe that the true geometry of space corresponds, not to $\{\hat{\Gamma}^a_{bc}\}$ and the metric $\hat{g}_{\mu\nu}$ ($\mu, \nu \in \{1, 2, 3, 4\}$), but to the alternative $\{\tilde{\Gamma}^a_{bc}\}$. Now, the actual observed geodesics are *inconsistent* with those predicted by

$$\frac{d^2 x^a}{dt^2} + \tilde{\Gamma}^a_{bc} \frac{dx^b}{dt} \frac{dx^c}{dt} = 0. \tag{2.2}$$

Our researchers naturally surmise that the observed trajectories are not those of particles in free fall, but are distorted by an extra force field. In the presence of a force F^a, say, the Newtonian equation of motion is

$$\frac{d^2 x^a}{dt^2} + \tilde{\Gamma}^a_{bc} \frac{dx^b}{dt} \frac{dx^c}{dt} = F^a. \tag{2.3}$$

Since (2.1), with $\hat{\Gamma}^a_{bc}$, describes the particle motion *empirically correctly*, the force is

$$F^a = (\tilde{\Gamma}^a_{bc} - \hat{\Gamma}^a_{bc}) \frac{dx^b}{dt} \frac{dx^c}{dt}. \tag{2.4}$$

F is a "universal" force. In general the force will not vanish at infinity, an extraordinary feature like its unusual dependence on the particle momentum instead of the particle's position. Those features apart there is little to guide our choice. However, as noted a few paragraphs back, the *conjunction* (unification) of Newtonian dynamics and classical electrodynamics singles out exactly one pair of Euclidean metric and (zero) "universal" force.[24]

Two distinct formulations of Newtonian gravitation in the geometric language are often cited in the literature as examples of (deductively) observationally indistinguishable theories, illustrating the (strong) "underdetermination of affine geometry in the context of classical physics" (Glymour, 1980b, p. 358). As a point of reference recall that a standard formulation of Newtonian gravitation theory has the following features: Galilean coordinates x^1, x^2, x^3 and universal time t are measured by "ideal" clocks and unit rods. The gravitational force is derived from a gravitational potential $\phi(x^1, x^2, x^3, t)$, which is connected to the mass density $\rho(x^1, x^2, x^3, t)$ by *Poisson*'s equation, and Newton's *Lex Secunda* takes the form $\ddot{x}^i = -\frac{\partial \phi}{\partial x^i}$ were $i = 1, 2, 3$.[25]

Formulation I: Newtonian space-time is a four-dimensional manifold equipped with a symmetric affine connection (derivative ∇) and a function t called "universal time" which is measured by "ideal clocks". In addition there exists a symmetric contravariant tensor h_{ab} with signature $(+ + +0)$ such that $h^{ab}t_b = 0$. The tensor field h defines the spatial metric; length calculated from h are measured by "ideal rods". (The space-time metric is degenerated.) The trajectories of free falling particles are described by the geodesic equation associated with the connection, see equation (2.1). The geometric field equations are summarized by:

---

[24]I am treating familiar turf, the exposition follows Trautmann (1965, pp. 417–420) and Glymour (1980b, p. 366).
[25]Adapted from Trautmann (1964); an alternative, rigorous axiomatization is developed in Misner et al. (1973, pp. 300–302).

Example 2.6.1

$$\nabla_c h^{ab} = 0 \quad (2.5)$$
$$\nabla_a t_b = 0 \quad (2.6)$$
$$t_{[e} R^a{}_{b]cd} = 0 \quad (2.7)$$
$$h^{ad} R^b{}_{cde} + h^{bd} R^a{}_{edc} = 0 \quad (2.8)$$
$$R_{ab} = -4\pi k \rho t_a t_b \quad (2.9)$$

(Square brackets denote anti-symmetrization in the indices; ρ is the mass density; indices range from 1 to 4; Einstein's summation convention is adopted; k is the gravitational constant; $R_{ab} = R^d{}_{abd}$.) Equation (2.9) corresponds to *Poisson's* equation, relating derivatives of the gravitational potential to the mass density.

Although each "slice" of space-time at fixed universal time is Euclidean, the whole space-time according to I is curved and the connection is *not flat*. Formulation I of Newtonian gravitational theory is provable (deductively) observationally *indistinguishable* from Newtonian mechanics as standardly formulated. How do they differ in point of physical ontology? The gravitational potential ϕ does not explicitly appear in formulation I, hence the theory now seems to be "committed" to an ontology with a *smaller number* of physical objects (provided the potential is not considered a mathematical fiction for calculating the force). On the other hand, the affine connection behaves like a physical object (the set of free falling particle trajectories determines the connection completely). Moreover, the geometric field equations entail the existence of a scalar function which has all the properties and relations of the gravitational potential ϕ. Note the increased number of "geometrical objects" in I when compared to the standard formulation.

Formulation II. This account of Newtonian gravitational theory is set up so that the field equations permit an inertial coordinate system in which the connection appears as *flat* (in components: $\Gamma^a_{bc} = 0$). With assumptions about the fields t, h and constants as before, the geometric postulates for the field equations are:

Example 2.6.2

$$\nabla_c h^{ab} = 0 \qquad (2.10)$$
$$\nabla_a t_b = 0 \qquad (2.11)$$
$$R^a{}_{bcd} = 0 \qquad (2.12)$$
$$h^{ab}\nabla_a \nabla_b \phi = 4\pi k \rho \qquad (2.13)$$

The trajectories of free falling particles are governed by a generalization of Newton's Second Law:

$$\frac{d^2 x^a}{dt^2} + \tilde{\Gamma}^a_{bc} \frac{dx^b}{dt}\frac{dx^c}{dt} = -h^{ad}\nabla_d \phi \qquad (2.14)$$

This formulation is provable (deductively) observationally indistinguishable from the standard formulation of Newtonian mechanics and observationally equivalent to formulation I. In a suitable coordinate system one can switch back and forth between the two formulations on account of the equation:

$$\Gamma^{IIa}_{bc} = \Gamma^{Ia}_{bc} - h^{ad}\frac{\partial \phi}{\partial x^d} t_b t_c.$$

How do I and II differ in point of physical ontology? Formulation II appears to be committed to (or postulates) the existence of a gravitational potential, separate from and beside the affine connection, if one grants the latter status as a physical object. Hence, there is an "ontic" difference. However, a word of caution is in order: an axiom system divides the theory's concepts into those, which are primitive and those, which can be derived from or reduced to the primitive ones. The induced division between "primitive" and "derived" need not coincide with the division into what is considered "real" and what is a mere mathematical auxiliary or fictive in the theory. (Does not the latter guide the construction of a system of axioms, or choice between various systems?) In a physical theory the (finite) set of primitive concepts includes physical quantities and objects as well as proper mathematical objects. In classical physics, for instance, it is possible to generally separate the (primitive) geometrical entities from the (equally primitive) physical ones. Yet, should the (primitive) metric tensor, the curvature, the connection or some traces and contractions thereof in general relativity count as physical objects? The proper criteria, I think, include both considerations of which objects are empirically measurable (and how, compare chapter 3), and considerations of symmetries intrinsic to a theory. For instance, one reason for why the vector and scalar potential of classical electro-dynamics were considered for a long time a mere mathematical convenience (in contradistinction to the electro-magnetic fields), was that these potentials can be nearly arbitrarily ("gauge") transformed without any consequences for the theory's predictions (see p. 73). The discovery of the *Aharonov–Bohm effect* (1959), an observable effect of the vector potential on trajectories of charged particles governed by the Schrödinger equation in an external electro-magnetic field, changed all this and restored physical significance to the vector potential. (This is another instance of the "conjunction option" in lifting certain types of underdetermination.) Or consider the criterion that objects count as truly physical and 'real' if and only if they have properties like charge, energy, momentum, etc., which are conserved when the total system of objects (in a model) is taken into account. The requirement is plausible, it accords fairly well with practise in physical disciplines. Since conservation laws themselves arise from symmetries (for instance, in classical mechanics conservation of momentum arises from translation invariance), the criterion for the 'ontological import' of a theory is not independent from wider symmetry properties of that theory.

Chapter 3
Rationality, Method, and Evidence

The deeper, explanatory and general features of our world are not accessible by direct observation. They require indirect epistemic access by way of a method of justification for our beliefs with respect to those explanatory features, or a method of confirmation for theoretical statements. Perhaps then empirical irresolvable conflicts of assertion, which are by necessity about those recondite, theoretical features of the world, can be traced back to relying on a faulty, "gullible" method of confirmation? This is what critics charge and the finger of blame has come to rest on a classical popular method of validation of theoretical beliefs: hypothetico-deductivism, or deductivism for short.[1] Deductivism, in one of its many variants, holds indeed a central place in Quine's epistemology. The present chapter examines this criticism along with a selection of alternative accounts designed to replace or significantly complement hypothetico-deductivism. I present reasons for why these accounts, promising as some of its results are, offer no general solution to the problems stemming from the underdetermination thesis at the present.

3.1 Deductivism Revisited

Deductivism underlies much of everyday judgements and belief formation, though rarely in an explicit form. Deductivistic modes of justification come to the fore in the context of controlled observation and

---

[1]For instance: "A tacit acceptance of this essentially inadequate account of confirmation [hypothetico-deductive confirmation – T.B.] underpins much of the favourite treatment of the underdetermination thesis" (Norton, 1994, p. 8).

test, particularly in those sciences that boast postulates, laws, and a formalized language. Its merits then are best discussed in the context of the exact sciences, to which I turn now.

Hypothetico-deductive methodology comes in many flavors: there is the classical account codified by John Herschel, William Whewell and John Stuart Mill; there is falsificationism, naive and sophisticated; there is Demonstrative Induction. In whatever form, there are few scientists throughout history who do not appeal to one version or the other in arguing their claims. To cite just one example from many in order to counter-balance the criticism that will follow. Huyghens studied the characteristic double refraction pattern of a beam of light transversing a calcite plate (the transparent Island Spar). When a beam of light is made to transverse two consecutive calcite plates it exhibits complicated empirical regularities, which were largely unknown at his time.[2] Huyghens succeeded in deducing the characteristic refraction pattern from the hypothesis that light propagates as a transverse wave through the ether in matter-free regions and extra geometric assumptions about the wave's propagation in the crystal and and the shape of its molecules. No other way of explaining the phenomenon in a similar systematic fashion was known at the time; its main competitor was Descartes' theory of refraction. In his *Traité de la Lumière* Huyghens plausibly claimed that the deduction *supports* the wave theory of light. Many other instances of hypothetico-deductive modes of justification in the history of the sciences can be adduced, often in the form of successful novel predictions, but this one may suffice.

Assuming a standard conception of the language of science, hypothetico-deductive (hd-)methodology for a deterministic theory or hypothesis consists of three elements: (a) the formulation of a hypothesis H typically involving terms not found in the observation vocabulary; (b) the deduction of testable consequences from the *explanans* H, true observational premises (initial conditions) and established auxiliary assumptions or independent background-information; (c) an *inductive rule*: if consequences derived according to (b) are verified than H is indirectly supported or confirmed; otherwise H is (incrementally)

---

[2]The exception is a work by one Erasmus Bartolinus, see Duhem (1978, pp. 40f).

disconfirmed. Suppose the set of all available relevant data (at a time t) for H takes the form of a set of n pairs of observation sentences O_1, O_2, where the first one specifies the initial condition. Schematically, to *hd-confirm* a (deterministic) hypothesis H relative to background knowledge B it is necessary that there is an n such that following conditions hold for each of the n pairs[3]:

HD 1   1.1 $B \& H \& O_1 \to O_2$

        1.2 $B \& O_1 \not\to O_2$.

The first requirement states that each O_2 can be derived or explained ('predicted') from the hypothesis together with the initial datum, given certain highly credible background assumptions. The second requirement states that the sentences in B alone do not suffice to derive or explain the observation expressed by O_2 given O_1, so that H is not superfluous, genuinely extends the content of B, and clearly 'stands out'. Note the dependence on n, the finite size of the total set of relevant data. It is asserted that if H is confirmed, then there is an n, small or (more likely) large, such that the two conditions hold on the set of relevant data. If n is small the conditions are too easily met, but the number may vary from scientist to scientist. The assumption that the total set of relevant data is finite is simplifying and realistic.

The total evidence condition, i.e. the requirement that there is no strong evidential reason to believe H to be wrong, is obviously met (provided one excludes as evidence 'circumstantial evidence', 'gut feelings', analogies and the like). It is slightly more realistic to require of only $m < n$ pairs in the set of all relevant data that they satisfy the two conditions, while the remaining $n - m$ are either *prima facie* falsifying instances or are counted as 'inconclusive' predictions. The fact that a few derivations turn out to be wrong (relative to the background-information) need not *per se* cancel the confirmatory force of the rest of H's consequences. Confirmation may be linked to a hypothesis's

---

[3] In stating this and the following requirements I am treating familiar ground, see in particular the discussion of hd-methodology in P. Horwich's contribution to Earman (1983), Glymour (1980b) and earlier works by G. Schlesinger cited therein.

reliability by way of a simple 'phenomenological' counting of the hits and misses of H. Thus the ratio $\frac{n-m}{n}$ could serve as a crude measure of the *degree of support* of H by the total set of relevant data, and hence as a means to pass from a qualitative or classificatory concept of confirmation to a quantitative one.[4] Provided, of course, n is not inflated by, say, frequent repetition of one successful experiment. Diversity of evidence is decisive, but not easily captured syntactically (but see below).

The "observational hd-content" of a hypothesis H is naturally defined as the class of all (non-analytically) true observation sentences entailed by $H \& O_1 \& B$ for each verified initial datum O_1, and not entailed without H.[5] Another name for the observational (or factual) hd-content of H's is *truth content*; similarly, wrong predictions form H's falsity content. The requirement above demands that the falsity content of H be empty (i.e. $m = 0$).[6] A derived notion is the observational content of H *relative* to an auxiliary hypothesis A, defined as the class of (non-analytically) true observation sentences which are entailed by $H \& A \& O_1$ for all admissible initial data O_1, after Carnap (1963a, p. 962). Observational *hd-equivalence* may be relativized accordingly: two hypotheses H, H' in L are observationally hd-equivalent relative to A if H' has the same observational content relative to A as H. Therefore two distinct theories may agree in all predictions based

---

[4]The fraction is set to 0 by definition in case $n = 0$; the fraction itself is the simplest choice among several forms a support function in n, m can take.

[5]It is unnecessary to identify a theory's hd-content with the 'truth content', false predictions are part of a theory's observational content too. The pre-fix "hd-" also acts as a reminder that the true factual content of a theory may not always coincide with its deductive content. See p. 108 for Quine's closely related notion of content.

[6]Deductive content and hd-equivalence can be generalized to *identity of entailed probability distributions* for the various observables, as suggested in Sober (1996, p. 176). For a given theory the error distribution (uncertainty) in the *initial data*, for example, will be mapped into an error distribution in the value of a predicted quantity. In general, however, empirical hd-*equivalent* theories may give rise to non-equivalent error distributions, due to the different way the observable results are computed in each alternative. In the following I consider only the case of 'deterministic' theories and hypotheses.

on A, and yet not be observationally equivalent with respect to the shared auxiliary A'. A broader comparison of hd-contents may consist in drawing up a list of auxiliaries relative to which T and its rival T' have identical hd-content, and another list of those auxiliaries relative to which their respective contents diverge. Relative hd-equivalence fails to obtain if different auxiliary hypotheses are employed in connection with H or H' for prediction and explanation, or if the hypotheses are formulated in non-equivalent observation languages.

Relative hd-equivalence is too strict a notion, for two hypotheses may be hd-equivalent without being A-relative equivalent. There is no particular reason to restrict hd-equivalence between hypotheses (or theories) to sameness of truth content with respect to a shared set of auxiliaries. The definition of the (relative) hd-content of a hypothesis appears to be *vacuous* if *every* auxiliary hypothesis is admissible (except when the hypothesis in question is not in need of auxiliaries), for every observation sentence becomes derivable from the hypothesis in question. J. Leplin has recently drawn attention to this uncomfortable fact in order to question certain arguments for underdetermination (see below).

Although the account given of hd-methodology so far is extremely rudimentary it appears to suffice to indicate why hd-methodology is inadequate and how it underpins underdetermination theses. Among its consequences is the following: if an hd-confirmed hypothesis H is entailed by another more comprehensive hypothesis, which is consistent with the initial data in the set of relevant tests of H and with the same background-information, the latter hypothesis is hd-confirmed as well: confirmation "flows" upward the entailment chain. Hempel's *converse consequence condition* is met, and thus the door is already open for a systematic form of underdetermination – namely one generated by *irrelevant* conjunctions to a testable hypotheses. Conjunctions are irrelevant if they do not contribute to the explanation of the data and are observationally isolated, typical examples include untestable 'metaphysical' claims. Thus every statement, which entails the hd-confirmed hypothesis, is itself confirmed: an absurdly holistic result. Once the qualitative concept of hd-confirmation is taken into account, the picture improves modestly. The degree of support of the more comprehensive hypothesis, generally, is a complicated function of (n, m) for each entailed sub-hypothesis including H. Interestingly however, the comprehensive hypothesis' degree of support may be *less* than the degree of support for H alone, as is easily seen in case where the comprehensive hypothesis is a conjunction of H with a separately tested hypothesis. Support "flows" upward, but it is attenuated if certain conditions are met. For a conjunction of an untestable isolated claim with an hd-confirmed hypothesis these conditions are not met, unfortunately, and the conjunction has the same degree of support than the actual testable *explanans* alone. Similarly, a (consistent) set of independent hypotheses is hd-confirmed as a whole, if one of its members is confirmed, and a theory is thus confirmed, if any sub-theory of it is hd-confirmed. These are instances of an *intrinsic confirmational indeterminacy* of the hd-account.

One proposal is to add a requirement to the effect that the hypothesis H cannot be written as a non-trivial conjunct of two sub-hypotheses, such that exactly one of the conjuncts satisfies the initial requirements for *hd*-confirmation. Apart from being too limited, at least in one formulation this suggestion encounters several difficulties (Glymour, 1980a). An alternative and more generally applicable, time-honoured proposal is an appeal to the methodology of "simplicity", "conceptual economy" or Ockham's razor.

An important manifestation of the confirmational indeterminacy of hd-confirmation is the well-known difficulty to identify the faulty hypothesis from a set of hypotheses, which collectively entail a prediction that has been falsified. *Duhem* showed in a number of detailed case studies that in sufficiently complex theoretical systems there is frequently an alternative derivation of the actual observation sentence, which allows to "safeguard" or "immunize" a salient conjunct hypothesis (for instance, the corpuscular hypothesis of light as the backbone of an explanation). Thus if the hd-account of confirmation is all there is to confirmation than by way of Duhem's thesis the existence of inferentially indistinguishable conflicting theoretical systems is made highly likely. I will inspect this argument below in more detail.

The litany of woes has not yet come to an end. It does not follow from the account so far that every consequence of an hd-confirmed hypothesis is hd-confirmed by the same total set of relevant data: a consequence of H need not by itself satisfy the first condition; only those sub-hypotheses of H that do are confirmed. This is troublesome, since the confirmation of a hypothesis should inspire confidence that some of its consequences, for instance, predictions beyond the set of all relevant data at time t, turn out to be true. It is tempting to repair the defect by requiring that every consequence of an hd-confirmed hypothesis is confirmed as well. But then any of the isolated, idle claims alluded to in the paragraphs above is hd-confirmed too, by itself and not as part of a statement package.[7]

A difficulty more easily overcome is that all n (or $n - m$) verified derived observation sentences contribute *equally* to the confirmation of H. It is a difficulty, since the observation of another falling leaf, or the successful use of the "Geopositional System" of satellites for navigational purposes hour by hour do not, it appears, hd-confirm general relativity.[8] One natural suggestion is that the *predictions be novel* in order to confirm a hypothesis. Novelty of an observation has many senses: ranging from 'not previously known' to 'not used in the formation of the theory'. A successful novel prediction or derivation is one, roughly, which is verified against well founded *expectations* (based perhaps on a competing hypothesis), or one regarding a subject area were none was possible (thinkable) before. To include the intuition that surprising or novel predictions are confirming one may add a requirement to the effect that our background information B by itself would lead one to expect O_3 instead of O_2, the two observation sentences being incompatible.[9] Thus, for instance:

HD 2 There at least $k \leq n - m$ predictions from H together with B such that, if O_1 is an initial datum and O_2 the conclusion of one

---

[7]The *locus classicus* for the difficulties connected with adopting the (special) "consequence condition" is Hempel (1966, pp. 30f).

[8]The point that confirming evidence is altogether too easy to produce has been emphasized by Popper; concepts of 'novelty' have been investigated recently in Leplin and Laudan (1993), Mayo (1996) and Leplin (1997a).

[9]Instead of keying novelty to highly credible, uncontroversial background information B, one may prefer to link it to what a strong *rival* hypothesis to H would lead us to expect (P. Achinstein).

of them, there is in each case a further observation sentence O_3 and background information B', satisfying the following conditions:

2.1 B' & $O_1 \rightarrow O_3$

2.2 $\neg(O_2 \& O_3)$

Certainly the strength of the deductive support will increase with k, although not necessarily linearly, but there is no need and perhaps no possibility to fix the value of the parameter k in advance. If only 'surprising' predictions support a hypothesis and $m = 0$, then the ratio $\frac{k}{n}$, or a more complex function in k, m, n in case $m > 0$, could serve as a measure of the degree of support of H by the total set of relevant data, instead of the original proposal $\frac{n-m}{n}$. In any case, my intention is to remind the reader that the present difficulty can be remedied; I do not wish to take a stand here on whether a requirement as such is necessary or sufficient for hd-confirmation, or rather belongs to the psychology of the scientist.

Hd-confirmation accrued by successful, perhaps novel predictions or explanations, is devalued, if not outright cancelled, if there is a known alternative hypothesis that matches these successes. This intuition seems wrong to scientific practice without inserting qualifications like 'credible' alternative hypothesis, and wrong from the point of view of a bootstrap condition on testing, yet it reflects the exclusive austere deductive relationships, which mark an hd-account of confirmation. In a sense, hd-confirmation accrues to a hypothesis provided it is the 'best' available deductive systematization of the facts. Those who share this intuition, will find the following requirement necessary for confirmation:

HD 3 For every hypothesis H, which satisfies the previous requirements, there is *no known* hypothesis H^* of which the following is true:

3.1 $\neg(H \& H^*)$

3.2 H^* and B are consistent

3.3 Conditions HD1 and HD2 above hold for H^* with $n^* = n$ $m^* = m$ $k^* = k$ on the same set of relevant data[10]

For instance, for J. S. Mill, and many of his contemporaries, the inverse square law of gravitation had no known empirically adequate rival and hence could pick up confirmation. The condition, however, cannot be met without further qualifications. For there are *known* rival hypotheses in the form of algorithmically constructed, 'parasitic' hypotheses, which meet the conditions, at least if we follow Kukla (1996). One may object that such theoretical artifices are not consistent with our background information B. Granted this is the case Kukla might want to question either the epistemic status of B, or the requirement itself. It is natural to try to exclude those 'parasitic' hypotheses by indicating them as not genuine scientific hypotheses at all, and modify the requirement accordingly. For the purpose of the present discussion I will disregard this particular problem.

This conventional sketch of hd-methodology in hand, it is time to assess the overall argument, which claims: hd-methodology is the reason to believe in the validity of the strong underdetermination thesis, and with full recognition of the essential shortcomings of the former falls any rationale for believing in the latter. I think the premise is doubly wrong. Hd-methodology has neither been shown to be "essentially" inadequate nor is it the only or main reason for believing in strong underdetermination.

To begin with the second claim. Hd-confirmation harbours three, in the present context relevant *prima facie* difficulties: (i) confirmation of conjunctions with isolated sentences; (ii) confirmational holism, or the lack of discrimination in a set of hypothesis, which together entail a verified observation sentence; (iii) the lack of discrimination in a set of hypothesis, which together entail a false prediction. With regard to the first point, it is not confirmation theory's *task* to select the scientific language, or to demarcate admissible from non-admissible predicates, or sequences of quantifiers, in the formulation of hypotheses. *Prior* decisions like these suffice to get rid of those unwanted conjuncts. Inductive methods, and the concept of evidential relevance,

---

[10] The requirement is too strict in some ways, and can be easily adjusted.

come into play on the basis of these and other basic *prior* decisions. To think that confirmation theory can settle all these questions and justify those decisions appears to descend from a verificationism like the one Ayer once defended. Thus hd-methodology does not give automatically license to all kinds of 'metaphysical' claims. A restriction to the effect that only those conjuncts of H are co-confirmed, which figure essentially in the derivation of a verified observation sentence, excludes most remaining kinds of irrelevant conjuncts. With regard to (ii), scientists may parse a sentential compound, hd-confirmed as a whole, more finely than the basic hd-account allows for. Yet, few have maintained this amounts to an 'essential' shortcoming of the latter, nor does it seem impossible to overcome.

The one serious sort of confirmational indeterminacy is the one mentioned in (iii). There are two parts to consider. The *modus tollens* conclusion that the conjunction of the hypotheses in the compound is false is un-objectionable. But why should one require more information or finer-grained evaluation from an account of confirmation than deductive logic permits in this case? The difficulty stems rather from the second part of (iii), namely that without such a finer-grained evidential evaluation in place all choices of a replacement for the falsified sentential compound (the theory) are equally hd-confirmed and rational, as long as they explain the observation. I am not sure that this conclusion by itself is objectionable; in any case, it follows from a rudimentary and incomplete account of hd-confirmation. What seems objectionable is the further claim that any hypothesis can now be upheld "come what may". Theory thus seems to be cut loose from its moorings in observation, and this is a reason to believe in strong underdetermination. However, the claim is not a necessary consequence of our account of hd-confirmation. It rather depends on an extra assumption about the availability of suitable auxiliary hypotheses (A. Grünbaum). Hence, it is not correct to argue that strong underdetermination springs from hd-confirmation; one might as well say it has its origin in deductive logic. I return to the Duhem–Quine thesis as an argument for the underdetermination thesis in the next section.

With confirmational indeterminacy reigned in, hd-confirmation theory does not appear essentially inadequate.[11] More likely it is in need of further development and deepening, which may occur along three dimensions: (1) The concept of hd-confirmation is perhaps better represented as an essentially *comparative* notion: it lays down conditions for when a hypothesis is more likely to be true than a serious rival hypothesis. The existence of a known rival (occasionally in the form of the "null-hypothesis") goes some way to illuminate the notion of evidential relevance. It also sheds light on the data themselves, how they are produced (outcome of severe tests, novelty, etc.). In this sense, the necessary condition HD 1, the starting point of our discussion, is defective. (2) The concept of hd-confirmation should be associated with a quantitative degree of support, perhaps in a 'reliabilist' vein indicated above. (3) The concept of hd-confirmation is better viewed as an umbrella-term: it may include requirements regarding the observational determinability of parameters of a theory (its potentials, forces, masses, etc.). The 'bootstrap' account has shown the way here, but it is misleading to think of 'bootstrapping' as an account of confirmation in its own right (see below). It demonstrates rather the intricacies and the potential of a more sophisticated hd-confirmation theory.

I like to return to an issue I have left open a few paragraphs above. The underdetermination thesis refers to the *observational equivalence* of two or theories, and hence to the notion of a theory's *observational content*. And here lies the rub. In the hd-framework, sketched above, the observational content is identified as the class of all (non-analytically) true observation sentences entailed by $T\&O_1$ for verified initial data O_1 (see p. 92).[12] This is a caricature. A theory's ability to predict or explain a fact depends on more than initial data, or even initial conditions: parameters have to be given values; potentials and force-functions have to be put into the dynamical equations; broader theoretical hypotheses act as selectors from a plenitude of solutions for

---

[11]Among the difficulties I have neglected – see the discussion in Laudan (1996, pp. 64f) – is that the set of evidentially relevant data for a deterministic theory may *overlap* with the deductive hd-content of a theory, and not strictly coincide with it. I address this issue in section 3.5.

[12]The dependence of the observational hd-content on background information (B) has been dropped, and reference to hypotheses has been shifted to theories.

those equation, etc. (It is formally possible to bring these considerations into the fold of the content formula given, but only at the price of 'inflating' the theory.) Moreover, nothing has been said about what to count as an observation predicate and observation sentence, i.e. about criteria for selecting the proper observation language. Carnap has favoured a universal physicalistic language and identified observation sentences as (non-analytic) singular sentences in the observation vocabulary. The difficulties for his position need not be rehearsed here.

One proposal for taking proper account of auxiliary hypotheses (A) is the notion of *relative* content of a theory T, the class of (true) observation sentences, which are entailed by $T \& A \& O_1$ for each initial datum O_1. Building on this notion one can define A-relative equivalence of two or more otherwise distinct theories. The definition as it stands is not very useful, since any sundry observation sentence (in a given 'observation language') can be made an element of T's relative observational content by a suitable selection of auxiliary hypotheses (Leplin, 1997a, p. 155). Carnap's hedge is the word 'admissible', but consistency requirements apart, which hypothesis is to count as admissible?

J. Leplin recently identified a more profound difficulty for the underdetermination thesis behind this question (Leplin, 1997a, pp. 154f). Many (potential) auxiliary hypotheses to a given theory, which help to enable predictions, are themselves purely 'theoretical' statements. On what basis is one justified in picking one over another – *given* the truth of the underdetermination thesis? If A is such an hypothesis, then in general there is no guaranty that any of the 'equally justified' alternative hypotheses to A, presumed to exist, gives rise to the same range of predictions in conjunction with T. Since the underdetermination thesis robs us of any reason to choose rationally among auxiliaries, it robs us of the basic notion of 'equality of observational content' as well. Moreover, the general availability of rival theories to T with *equal* observational content was the very premise for inferring the underdetermination thesis in the first place: '$EE \rightarrow UD$' in Leplin's terminology (see section 1.1).

In reply, one should point out that the 'underdetermination thesis' *does not apply* to theoretical auxiliary hypotheses. The reason is that theoretical auxiliaries are mini-theories with *insufficient* 'critical mass' or scope. For instance, the Pauli exclusion principle is a statement about the anti-symmetric form of the quantum-mechanical state of a finite system of identical electrons. The hypothesis is crucial for the successful application of quantum mechanics to the structure of atoms and their chemical properties. Pauli's hypothesis has no intrinsic independent observational hd-content and no support from background theories apart from quantum mechanics. One factor for why scientists believe it is true, beside its simplicity, is that it enables uniformly correct explanations in conjunction with the postulates of quantum mechanics. The latter have many diverse applications, apart from elucidating the structure of atoms; Pauli's principle does not and it makes hardly sense independent of quantum mechanics. Thesis U, one expression of the thesis of strong underdetermination (p. 19), does restrict its claim to theories of 'sufficient critical mass or scope'. It appears immune to Leplin's argument.[13]

Is it true that a theoretical auxiliary hypothesis is admissible only if scientists "affirm" the hypothesis independently from its potential deployment in conjunction with a specific theory? In particular, does it have to be better supported than the theory in question (Leplin, 1997a, p. 156)? This is a bone of contention. Leplin points out that Lorentz' electron theory depends for its observational equivalence with special relativity theory on certain auxiliary hypotheses, which *lacked* independent support, suggesting this may be one factor for why Einstein's 'program' eventually superseded Lorentz's (Leplin, 2000, p. 380). As an initial assessment this observation is true enough. However, suppose Lorentz's theory would not have had a known rival at the time. It does not take a great stretch of imagination to see that over time those obscure auxiliaries would have shared the support the theory derived from its successes in explanation and prognosis. (A *Bayesian* analysis of this case bears this out, I believe.) The epistemic status of Pauli's principle with respect to quantum theory is similar.

---

[13]Note, Leplin's argument does not threaten the notion of relative hd-equivalence of content, since theories are compared with regard to predictions based on a shared auxiliary hypothesis – without regard to its credentials.

Leplin rejects the idea that auxiliaries might be selected instead according to their 'conformity to observation' because this criterion is too permissive. Nor can admissibility be explicated as meeting standards of 'acceptability' in the sense of constructive empiricism (I. Douven), because acceptability claims are no substitute for the presumption of truth, and are too weak to determine a theory's content uniquely (Leplin, 2000). I am not persuaded by these rebuttals. The reason is that I do not think that the observational content of a theory is solely determined by the theory. In general there is no such thing as a "commitment" of a theory to certain observational consequences (compare section 3.5).

3.2 Quine on Method and Evidence

Quine has been charged with recklessly rejecting methodological canons and normative considerations in theory choice, and with opening the door to relativism (see Laudan, 1996). The charge is unwarranted:

> All this leaves the heuristics of hypothesis untouched: that is, the technology of framing hypotheses worth testing. This is the domain of [...] Ockham's razor. It is the domain also of standard deviation, probable error, and whatever else goes into sophisticated statistical method. (Quine, 1991, p. 274)

All in all, Quine adhered to the hypothetico-deductive approach, reconfigured in a naturalistic vein. He drew a distinction between the point of view of scientific practice (where pragmatism and probabilities reign) in matters methodological and the point of view of epistemology (where deductivism reigns): a relationship not unlike that between civil engineering and particle physics.

His foundational considerations are governed by the idea that deductive logic is the "mother-tongue of science". Deductive logic alone relates theory to experience and observation in a rational reconstruction of science. Observation sentences can refute a cluster of hypotheses by way of implication (*modus tollens*); they cannot verify or confirm a hypothesis. This apparent commitment to austere falsificationism in methodology and epistemology, though, is incomplete. In some passages Quine wrote as if observation sentences function not solely as refuting instance or as "checkpoints" but can positively raise the likelihood of *truth* of a theory to the point of verification. There is nothing wrong, according to Quine, in holding that our *total theory*, the claims of the sciences and of common sense taken together (where they do not conflict), is *verified* if the total theory is in agreement with "all" evidence. Provided, of course, there is no known incompatible and untranslatable rival total theory – a possibility Quine has discussed in a number of places (see 5.1, 6.5). Quine also rejected the Popperian view that we can lay down rules for the game of science ("critical", "progressive", etc.) in an *a priori* manner and hold scientific practice up to this measure. For the naturalist, as well as for thinkers with a historicist bent, this relation between scientific methodology and scientific practice is reversed, or rather a dialectic relationship. Moreover, strict falsificationist methodology is conventionalist with regard to the "Basissprache", i.e. what *kind* of sentences count as potential falsifiers. Quineian naturalism locates observation sentences where and when the human observer utters them. Observation sentences are, roughly, sentences which all competent speakers spontaneously and uniformly assent to or dissent from when exposed to corresponding stimuli under sufficiently similar conditions. There is no alternative observation language to "choose". (Observation sentences on this account need not be true.)

With regard to the practice of natural science Quine held that mainly the "maxims" of global simplicity and conservativeness in revising our conceptual system govern scientific methodology in the face of adverse observations and temper falsificationism. *Relative* to accepted background information probabilistic reasoning and methodological rules become effective and decisive in preferring one theory over its alternatives. The "technology of anticipating sensory stimulation" – relative to accepted background information – can even sometimes *resolve* underdetermination of pairs of equivalent theories. Ampliative inference and induction play a role, and an important one of course, but they are highly conditioned on human cognitive capacities as a product of evolution and emerge from a contingent history of trial and error learning in matters methodological.

It is not the intentional indeterminateness and generality of his statements on hd-methodology that has attracted most objections, but rather Quine's views on confirmational holism. Here, indeed, lies *one* reason for his commitment to the generality of underdetermination.

> This holism thesis [i.e. the Duhem–Quine thesis] lends credence to the under-determination thesis. If in the face of adverse observations we are free to choose among various adequate modifications of our theory, then presumably all possible observations are insufficient to determine theory uniquely. (Quine, 1975, p. 313)

As I have pointed out, to go from *modus tollens* holism in a set of hypotheses to any hypothesis can be held up 'come what may' and on to strong underdetermination claims requires extra steps. One of these steps is the assumption (i) that one cannot assert the truth or falsity of any of the conjunct hypothesis *independently* from the present (refuting) test, and independently from the remaining conjuncts. If all conjuncts were hd-confirmed and established independently from the "hypothesis under test" then falsification by a refuting pair of observation sentences (O_1, O_1) is rationally compelling. Suppose that in a specific theory the assumption holds. (ii) The second premise is the general *availability* of "auxiliary" hypotheses, which can replace one of the conjunct hypothesis, so that the new sentential compound includes the "hypothesis under test" and is (as a whole) hd-confirmed by the initially refuting datum. On the face of it, it seems extraordinarily *improbable* that (ii) holds in a given case. However, as is well-known, *Duhem* had found just that in a number of case studies. In sufficiently complex theoretical systems there is frequently an alternative derivation of the actual observation sentence, which allows to "safeguard" or even confirms a salient conjunct hypothesis (i.e. the corpuscular

hypothesis of light).[14] Thus if the initial pair of observation sentences is refuting
$$(B\&H\&H'\&O_1 \to O_2)\&\neg O_2$$
The crucial claim can be rendered schematically as
$$\exists H''(B\&H\&H''\&O_1 \to O_3)$$
where O_3 is the actual observation sentence.[15]

Now, to underwrite strong underdetermination one would need a proof or at least a convincing argument for the truth of that (Quinean) claim. No such argument exists to my knowledge, and in its absence one can at best ascertain that such auxiliaries may or may not exist depending on the circumstances.[16] Once natural requirements on the admissibility of an alternative auxiliary H'' are imposed, the thesis is clearly implausible (Laudan, 1996, pp. 36f). For instance, one may require that the alternative auxiliary hypothesis is drawn from accepted background knowledge B, or that the hypothesis has certain formal

---

[14]The *locus classicus* for Duhem's observations is Duhem (1978) part II, chapter X. Quine's most radical affirmation of the thesis dates back "Two Dogmas": "The unit of empirical significance is the whole of science. [...] Any statement can be held true come what may, if we make drastic enough adjustments [...]." The following statement could have been made by Quine but actually is from H. Weyl's 1928 *Philosophie der Naturwissenschaft*: "Es ist sicher, daß ein beträchtlicher Teil des theoretischen Systems allen Erfahrungen gegen"über aufrechterhalten werden kann, wenn für den Rest Modifikationen gestattet sind" (Weyl, 1948, p. 116). Weyl replaced the dichotomy of a priori und a posteriori true, similar to Quine's move, by "Grade der Festigkeit" of hypotheses and first principles. Compare Carnap: "Widerspricht ein Satz, der L-Folge bestimmter P-Grundsätze [i.e laws of nature – T.B.] ist, einem als Protokollsatz aufgestellten Satz, so muß irgendeine Änderung des Systems vorgenommen werden: man kann z.B. die P-Bestimmungen so ändern, daß jene Grundsätze nicht mehr gültig sind; oder man nimmt den Protokollsatz nicht als gültig an; *man kann aber auch die bei der Deduktion verwendeten L-Bestimmungen ändern*. Es gibt keine festen Regeln darüber, welche Art der Änderung zu wählen ist" (Carnap, 1934, §82; my emphasis). In the same passage Carnap endorses hypothico-deductivism as the methodology of testing a hypothesis ("nachprüfen") and later *holism* (§82).

[15]Here I follow Grünbaum (1960, p. 118).

[16]Stegmüller et al. claimed to have a proof of the thesis. I have discussed the thesis from a (personalist) Bayesian point of view in Bonk (1994).

characteristics, etc. Indeed, Grünbaum constructs a counterexample to the thesis (Grünbaum, 1960, p. 119).

Quine has rejected the second implication above as an expression of the thesis and adopted a weaker, "legalistic" interpretation[17]: an *evidential holist* only needs to show that the hypothesis under test can be made logically *consistent* with the refuting observation pair. Consistency can be achieved simply by "dropping" (rescinding, "giving up") one of the other conjunct hypothesis necessary for the deduction of the wrong prediction O_2. Or more radically, by allowing for changes in the cognitive meaning of constituent terms of H, or by changing the logical and mathematical framework in which the theory is embedded. Quine concedes that Duhem's thesis is false for "practical purposes", i.e. not generally true given the strictures of scientific practice, but maintains that it is true for "philosophical purposes", i.e. from an epistemological point of view.

The proposal has two parts. First, in order to escape logical conflict with the recalcitrant datum and *modus tollens* one of the auxiliary hypothesis is 'dropped'. Second, at a later stage eventually an explanation of the datum on the basis of the 'saved' hypothesis is expected. Dropping an auxiliary hypothesis without replacement means shrinking the theory's observational hd-content.[18] If one is testing a theory against another theory, one that does explain the recalcitrant datum, then *ad-hoc* shrinking the former theory's observational hd-content is as confirmationally disastrous as a falsified prediction. A scientist is hardly justified in upholding the theory in those circumstances, or a crucial hypothesis as part of it, if an observationally adequate alternative is known. If no alternative is known, the credibility of the disabled theory will depend on how quickly a convincing explanation of the recalcitrant datum is forthcoming. This brings the second stage of Quine's proposal into play, the need to find a suitable auxiliary, which enables a correct explanation, and possibly allows for additional content. Yet, the very existence of an admissible auxiliary hypotheses

---

[17]See Quine (1962, p. 132; 1975, pp. 314–315).

[18]In case the suspect auxiliary hypothesis is independently well-confirmed then "dropping" it for no other reason than to "immunize" a part of the theory under test, would not count as properly justified.

is not universally certain. In neither case then do we have a strong reason to believe that any hypothesis can be made core of an observationally adequate theory. Hence this argument does not support a thesis of strong underdetermination.

In view of Quine's erstwhile claim that total science is underdetermined by all 'possible' observations, a few remarks about his notion of observation sentence and the meaning of 'all possible observations' are in order (see Quine, 1975). As mentioned briefly above, an observation sentence is a sentence that any witnessing, competent speaker in a language community is disposed to affirm spontaneously when queried. ("Occasion" sentences are learned by conditioning to direct stimulations caused by the event witnessed and guided by ostensive gestures. They need not command community wide assent.) Recourse to stimulus observability – as against aided observations, which may still count as observations – contributes to holism and the "general underdetermination" of natural knowledge. The epistemic distance it creates to the theoretical hypotheses under test, deeply embedded in the theory, gives rise to radical confirmational holism. Microscopes and telescopes, voltmeters, spectroscopes, computers, scintillation counters, etc., all the instruments the advanced sciences rely on to magnify effects and test theories, are just more theory between observation event and critical hypotheses – in the epistemological perspective. Stimulus observability is, the naturalist in matters epistemological maintains, the developmentally or genetically prior notion to a (i) theory-internally defined notion of observability and (ii) the experimentalists' and theoreticians' shifting criteria of "phenomenology". Theories that are indistinguishable relative to observations couched in the language of what scientists take phenomenology to be will be inferentially indistinguishable relative to the "lowest common denominator" sense of observability too. The reverse is in general not true. The naturalist's project is, after all, to account for the "projection" of the scientific worldview from the ground up, that is starting with the neural input on the human theorizer as the only source of information about her environment. Quine's notion is an *absolute* notion of observability in that it is maximally theory-independent and objective, although it is bound up in the shared language of the speech community and hence is theory-laden.

Quine defined a scientific theory's empirical content as the class of "observation categoricals", i.e. sentences of the form $O_1 \to O_2$, that are implied by the conjunction of the hypotheses of the theory together with background assumptions.[19] The premise O_1 describes the initial datum, and the categorical asserts that whatever is described by the two observation sentence will always occur together. Quine's admittedly schematic notion is equivalent to the "relativized" one defined earlier, as is easily shown by using the deduction theorem (compare p. 92), where it not for a difference in what the observation sentences are. For Quine O_1 could be something like "the needle of instrument no. 3 is now on the red mark", while in practice O_1 would typically be more like "a spin-up polarized neutron moves along the x-axis with p = 0.1 meV".[20]

Underdetermination of a theory, or of total science, by all "possible evidence" is a notion critics have justly found puzzling.[21] Intuitively speaking one wants to say that more than one theory, or one system of total science, each non-trivially different from the other, can account for any physically accessible fact one may discover. Ideal knowledge of all physical facts, were such omniscience possible, together with a standard methodological canon, cannot not uniquely fix the best theory, or total science, which accounts for them in hypothetico-deductive fashion. I review three explications of 'all possible evidence'.

(1) The first proposal is the identification of the totality in question with the totality of time-place indexed observation sentences. Supplementing observation sentences with the times and places of the (potential) observation act, relative to some fixed global coordinate system, generates a manifold of true or false "pegged" observation sentences.

---

[19] As the example in Quine (1990, p. 8) indicates Quine was aware of the necessity to take background assumptions into account.

[20] Quine's notion of observation categorical supplanted the earlier "pegged observation sentence" and "observation conditional" in Quine (1975, pp. 316–317). The notion of "reconstrual of predicates" too has disappeared from later writings on the matter (ibid. p. 320).

[21] In later discussions on related matters Quine pointed out, correctly, that it is sufficient for his purpose to consider empirical *equivalence* between rivals, not the existence of rivals in the problematic limit of all evidence.

3.2. Quine on Method and Evidence

The actual world picks out the correct valuation for these "atomic" sentences in a consistent way, irrespective of whether an observer with the disposition to assent or dissent when queried was there or (for the most part) not (Quine, 1970).[22] (Note, consistent valuations other than the actual one correspond to 'possible worlds', but surely not all are 'possible *experiential* worlds' for human observers. Suppose one can meaningfully delineate the latter among this totality. It is then tempting to define a strong notion of observational equivalence for rival systems of belief if and only if they agree not only in this world, but in all possible *experiential* worlds. Such a condition, if it obtains for two theories, would still not fully licence the inference to theoretical equivalence.) The relevant facts for the assessment of a theory include, on this account, those at remote times and places, past and future, which are physically inaccessible to humans. This expansion of observability beyond the contingencies of human history is necessary for a non-trivial notion of theoretical underdetermination. Yet, one wonders whether Quine's formulation is felicitous since evaluating the stimulus-defined *observation* sentence regarding the position of a certain atom 30 minutes after the Big Bang, say, does not seem to make clear sense. Another difficulty is that this formulation excludes from the relevant evidence: (a) accepted "theoretical" statements in physics, (b) the results of observations made by the use of measurement instruments,[23] and (c) counter-factual statements about what event would have happened had the physical circumstances of its presumed causes been different. Given Quine's extensionalism the latter restriction (c) is not surprising.[24] The first two restrictions reflect the central epistemic role Quine granted stimulus observation sentences in accounting for

---

[22] In view of Quine's later writings the observation sentence should be replaced by the observation categorical. For my purpose this refinement is of no consequence.

[23] The results are direct reports about the shape and number of cells, say, if the instrument is a light microscope. 'Cell' is not usually taken to be an observable predicate. Otherwise the deliverances of observations garnered in this fashion could be subsumed under the totality of "pegged" observation sentences.

[24] Putnam used contra-factual dependencies to generate systematically "alternatives" to the common sense world view. See section 4.4 for why I think this does not work.

the projection of our system of the world. However, the restrictions contradict sound scientific practice. Faced with this dilemma I think the course to take is to adjust our epistemology.

(2) One way to take "all physical facts" into account – not only all observable facts – is to specify basic physical predicates and to provide their truth value at every space-time point or event (Friedman, 1975). For instance, $P_{13}(x)$ may stand for 'proton at x' or a field-strength at x, where x ranges over points of the appropriate space-time manifold. Similarly for many-place predicates. The (naive) physicalist's claim is that ideal knowledge of all these values uniquely 'determines' what is true or real in the world. If there are two rival, global theories, and none is a mere notational variant of the other or differs on account of an inner symmetry, then one or both conflict with an atomic physical fact as specified above. If the two are observationally, deductively equivalent then the conflict will not have observable consequences – at least if we consider only this world among possible other worlds. Nevertheless, our ideal knowledge always yields a decision about which is the true one. Realistically, the totality of physical facts is specified by way of an accepted, global theory, or a finite, inter-related "patch-work" of theories. The basic predicates and corresponding atomic sentences are the ones of this theory or the patch-work, eventually after first performing a reconstruction of the theory in terms of first-order logic. If assignments of values to theoretical quantities differ between a primary and an underdetermined rival theory then the two theories are theoretically in-equivalent. Which is true and which false remains empirically unsettled (on a realist reading).

(3) According to a proposal in Wilson (1980) the relevant sense of observation is that of a *measurement*. The latter can be schematized as the interaction of two physical systems, the object system and the instrument, both described by the theory in question (this restricts these considerations to "complete" theories capable of describing their measurement instruments). The brief interaction correlates the physical states of the two systems. The outcome, as predicted by the theory, is a permanent change in the state of the instrument system, recordable as a meter reading q. Thus, a *possible observation* is represented as an ordered tuple $\langle S, \tilde{S}, q, t \rangle$, where, $S, \tilde{S}$ denote the two systems, or

models, or applications of T. The notion of 'observational equivalence' between two rival theories turns on matching every potential system or model under T with a model under the rival T' such that the latter is observationally indistinguishable from the former measurement interaction by interaction. Actually, we need not assume a global match between every potential model, but only a global map between actualized or realized models. If T explains a measurement event by the interaction of two systems, then the rival explains the same event and the same experimental outcome as the interaction of two systems, each of which is the image under the global map of the two primary systems.

None of these three notions is entirely satisfactory as characterization of "all possible observations" and observational equivalence. The first, the Quinean notion, is too overtly epistemic in nature and too removed from scientific practice; the second is problematic because it takes observationally inaccessible atomic facts into account. The third notion is problematic if only for the strong requirement that the secondary, alternative theory explains the (measurement) event in a structurally similar way as a measurement event between copies of the primary systems. There is simply no reason to suppose that, in general, the alternative theory describes the meter reading as the result of a measurement. Fortunately, one can explicate the notion in question indirectly by way of the familiar *experimentum crucis*. Two theories are not empirically equivalent if and only if there is an observation sentence that is entailed by one of the alternatives together with its proper auxiliary hypotheses (hence by an application, or model) and whose negation is entailed by the other (a rival model). If there is no such 'experiment' then the two theories are observationally equivalent. We do not have to decree anything about whether observation refers to the readings of measurement instruments or to Quinean observation categoricals, say. Nor, if only current instrumentation is permitted, or yet unimagined, future super-microscopes as well. Relevant is the existence of models (theory plus auxiliaries) that have conflicting observational consequences. The advantage of the present formulation is that we take the theories themselves, or generally systems of belief, as the guide for what to count as relevant evidence.

3.3 Instance Confirmation and Bootstrapping

In this section I examine the first of two recent proposals for a theory of confirmation, which aim to show how laws and values of non-observable quantities in a theory can be *deduced from the phenomena* and background theories. If these proposals are successful we have made a major step in remedying shortcomings of the hypothetico-deductive methodology and toward a satisfying resolution of empirical hd-indistinguishability between alternative theories. I argue that these proposals, *Demonstrative Induction* and "instance confirmation", fail to deliver on their larger promises.

In his attempt to find a precise definition of confirmation, and after he had rejected the "prediction criterion" of confirmation, i.e. a version of hypothetico-deductivism, C. G. Hempel developed a "satisfaction criterion" for the confirmation of a hypothesis by an observation report. According to the satisfaction criterion universal or existential hypotheses couched in first-order logic with one- or two-place predicates, are confirmed by an observation report if the finite number of individuals mentioned in the report satisfy the hypothesis, or rather its finite "development". The finite development of a hypothesis is, roughly, what the 'truncated' hypothesis asserts for a finite world consisting of just those objects mentioned in the observation report. Confirmation consists in demonstrating that the "development" of the hypothesis under test actually follows from the observation report (Hempel, 1958a). Clark Glymour shares with Hempel the intuition that a single instance can confirm a hypothesis – in the sense of a reason to believe it is true. A positive instance, typically, consists of a set of values for all the quantities (variables), which occur in the hypothesis under test such that the hypothesis is satisfied. The full set of values is, again typically, computed from measurements with the help of other hypotheses. The instance *confirms* the hypothesis only if certain conditions are satisfied: the computation is not circular; measurements could have led to a set of values that does not instantiate the theory, etc. Confirmation of H by data E is essentially relative to a body of theory, which in some cases may include the very hypothesis under test. The potential impact on the problem of empirically equivalent,

3.3. Instance Confirmation and Bootstrapping

conflicting alternative theories is obvious: for instance, although deductively hd-equivalent, one of the alternatives may have a theoretical structure that allows for instantiation in this sense, while the other does not. In this case, the former is "instance" confirmed and the other is not. Confirmation by satisfaction provides (1) a (novel) reason for rejecting theories which postulate "redundant" quantities, and offers (2) a method of differentiating between alternatives on the basis of the number of ways in which a hypothesis can be tested (or quantities can be measured): "the body of evidence that distinct theories hold in common [...] may nonetheless provide different support for the two theories, more reason to believe one than the other, more confirmation of one than the other" (Glymour, 1980b, p. 342). First, I will sketch this account of confirmation as satisfaction, and then I point out some difficulties with respect to theory comparison. A comprehensive evaluation of Glymour's interpretation of the satisfaction criterion is not my primary objective here.[25]

Systems of algebraic equations afford a simple exposition of the salient points of instance confirmation. Such systems may, for instance in the social sciences, represent statistical correlations between certain random variables and (not directly observable) quantities, which represent the "strength" of causal relations between those random variables.[26] Consider then two equational systems for six directly observable and measurable (independent) quantities a, b, c, d, e and f and four respectively six "non-observable" quantities x, x', x'', y, z and u. The set on the left represents $T1$, the one on the right $T2$:

---

[25]'Bootstrap-style' confirmation has drawn a lot a of criticism in form of counterexamples early on, to which Glymour responded by detailing a modified account in his essay in Earman (1983) and Earman and Glymour (1988). The modified account has been shown to have highly unintuitive consequences (Christensen, 1990); see also the critical essays by P. Horwich, B. van Fraassen and A. Eddidin in Earman (1983). I do not share the basic intuition of Hempel and Glymour that a single "instance" can confirm any hypothesis of an interesting degree of generality. Inductive learning does not appear to proceed in this fashion.

[26]I adapt the example from Glymour (1980b, pp. 298f); van Fraassen presents a somewhat similar reconstruction in van Fraassen (1980b, pp. 28–29) without the notions of computation and representative.

$$T1: \begin{cases} a &= xy \quad (3.1) \\ b &= xz \quad (3.2) \\ c &= xu \quad (3.3) \\ d &= yz \quad (3.4) \\ e &= yu \quad (3.5) \\ f &= zu \quad (3.6) \end{cases} \qquad T2 \;:\; T1 \,\&\, (x = x'x'')$$

Both theories entail two equations written only in terms of observable quantities:

$$af - be = 0 \qquad (3.7)$$
$$af - cd = 0 \qquad (3.8)$$

These two consequences exhaust the theories' *empirical hd-content*.[27] The second model $T2$ is obviously not a paragon of efficiency and theoretical elegance in its postulation of two extra non-observables. $T2$ would be rejected on the hypothetico-deductive model of justification if conjoined with either (1) a methodological rule of simplicity in equational systems, like Ockham's razor; or (2) with the requirement that all "parameters" be determinable from the data. Confirmation by instances is different in that it builds determinability of quantities into the very conditions for confirmation.

Theoretical quantities x, y, z, u in $T1$ are indirectly determinable experimentally if they have expressions by functions in terms of the observable variables only ("computations"). Computations depend on the truth of certain hypotheses, in this case drawn from $T1$. Substituting these computations for each theoretical quantity x, y, z, u into an equation or hypothesis $H(x, y, z, u)$ yields another equation $h(a, b, c, d, e, f)$. This equation h is a numerical relation between measured values, the representative of H, and varies with the hypotheses used in the computations. Suppose h does not reduce to a mathematical identity and does not depend only on a subset of the variables, so

---

[27] A third equation is listed in Glymour (1980b, p. 299), which follows from the two above by eliminating the product af.

3.3. Instance Confirmation and Bootstrapping

that its truth value varies in the 'right way' with the measured results.[28] Then it is natural to think of the measured values for a, b, c, d, e, f to have *tested* the hypothesis H – *relative* to the body of theory that made the computations possible in the first place.

For instance, the theoretical hypothesis (of $T1$) $a = xy$ has as its representative $af - be = 0$, computed with the help of A.2, A.5 and A.6, hypotheses drawn from $T1$. Thus a set of measurements may test *one* hypothesis from a conjunction of several.

Clearly, one set of equations is *better testable* by a set of data than a 'rival set', if more equations, which are not analytical truths, are testable in the former than in the latter; or if although all equations in both sets are testable, in the first set there are more independent ways of testing a given equations, than in the rival set.[29] With regard to the second equational set $T2$ values cannot be calculated in the required way that would instantiate, for example, the first equation. The obstacle, of course, are the "non-observable" quantities x' and x'' for which no value *individually* can be derived from any measurements. $T1$ is *better tested* than $T2$ (Glymour, 1980b, pp. 310–312). Since $T2$ is equivalent to $T1 \& (x = x'x'')$, the data may be said to test (instance confirm) a conjunct but do not test the whole conjunction.

The crucial notions developed here are applicable not only to social or physical systems, which are to be modelled by systems of equations. Glymour makes a case for that this pattern of testing and confirming by instances sheds light on methodologically relevant differences between Ptolemaic and Copernican theory, on certain claims and arguments used by Kepler and Newton, and on the differential value of tests of general relativity. Amongst the methodological benefits of confirmation by instances is, he claims, an explanation for the scientist's push for a variety of evidence in testing theories. In point of theory comparison consider the case of Newtonian mechanics formulated as a four-dimensional *space–time theory* (formulation A), see section 2.6 and Trautmann (1965, 1964), Friedman (1983, pp. 92–94). Particles are affected only by the gravitational potential which is determined

---

[28] For the detailed set of requirements see Glymour (1980b, p. 307).
[29] See Glymour (1980b, p. 310) for the complete set of requirements.

by the mass density through the spatial *Poisson equation*. The metric is flat and the Riemanian curvature vanishes identically. Motion of free particles in an inertial coordinate system is invariant under the *Galilei-group* of transformations of coordinates (this choice eliminates "absolute velocity" and the rest frame from the original Newtonian formulation). In a suitable coordinate system, if all the matter in the universe is concentrated in a finite region and if a boundary condition imposes that the potential vanishes at infinity the Poisson equation permits a *unique* solution for the gravitational potential (Friedman, 1983, p. 97). Under these global conditions our Newtonian field theory *has* a class of models in which the potential (and the connection) can be determined from the data: the trajectories of free falling bodies, etc. However, the determination is not possible for all types of models for the theory. Glymour concludes that *the theory is less well tested* than the hd-indistinguishable alternative (B) (Glymour, 1980b, p. 362): the equation of motion for the field theory formulation in the presence of the gravitational potential can be transformed (formally) into an equation which would describe the trajectories of free falling particles (the geodesics) in a curved four-dimensional space-time manifold. Details aside, the re-formulation makes it possible to "geometrize away" the gravitational potential by introducing a non-flat connection in a way that creates an *hd-indistinguishable* alternative to the version of Newtonian gravitation considered before. The upshot is that the same (infinite class of) observations on trajectories would permit technically the observational determination of the connection, and a test of a few of the postulates relative to the theory under consideration. The present formulation of Newtonian gravitation (B) postulates a smaller number of independent objects, and hence would be preferable on everybody's methodology.[30]

To return to our immediate question: it was the confirmational indeterminacy of hd-confirmation, which appears to engender a strong underdetermination thesis. How does the present account of testing

---

[30] Are formulations (A) and (B) of Newtonian gravitation "fully equivalent", since the second emerges as a trivial formal transformation of the first version? Glymour suggests the answer is No (Glymour, 1980b, p. 362; Friedman, 1983, p. 121, fn.).

3.3. Instance Confirmation and Bootstrapping 117

fare in this respect?[31] Suppose our data falsify equation A.7. Since the equation is a consequence of A.1, A.2, A.5 and A.6 together *modus tollens* permits the conclusion that one or more of the four is false, it does not tell which one it is. It is of no help to declare equation A.7. to be the representative of A.1, relative to 'computational subtheory' A.2, A.5 and A.6, so that the falsification is directed against A.1 alone. One may declare A.7. to be the representative of any of the other three equations with equal justification. At least, the sub-theory of $T1$, consisting of equations A.3 and A.4 is not thereby falsified, and indeterminacy is partially contained? The answer is yes. Yet partial detainment follows from a standard hd-account of confirmation as well, since the two equations are not essentially involved in the derivation of A.7. I conclude that in point of confirmational holism testing by generating instances is not, in general, superior to an orthodox hd-account of confirmation.[32]

A second source of systematic underdetermination of a theory, with respect to hd-confirmation by a positive test, is the extension of the theory by way of *irrelevant* conjuncts with isolated sentences, or by introducing gratuitous unobservable quantities. It seems that hd-confirmation, converse consequence condition in place, generally is unable to sort these cases out on the basis of evidence, while instance confirmation fares better. A simple example is theory $T2$, which replaces one variable in its close relative $T1$ by two in a way that turns the extra hypothesis $x = x'x''$ into an untestable claim.

'Conjunctivitis' is not, however, the only argument for strong, general underdetermination; for instance, it does not appear to have moved Quine. Removing this source of underdetermination does not funda-

---

[31]Glymour prefers the expression 'positive test' instead of 'is confirmed by data' E.

[32]If one rejects A.1 or A.6 on account of the falsifying datum, then the derivation of equation A.8 is blocked. A.3 and A.4 cease to be testable. Perhaps here is a reason to direct falsification towards A.2 or A.5 instead of A.1 or A.6? Note, equations A.7 and A.8, which summarize the theory's observational content, are formal *solvability conditions* for the system of equations $T1$. If satisfied they guarantee that the unknowns x, y, z, u have a unique value. This may be taken to indicate that A.7 and A.8 cannot 'represent' any particular hypothesis of $T1$.

mentally solve our problem.[33] But suppose it is the only or the best systematic argument. Traditional considerations of *simplicity*, attached to an account of hd-confirmation, remove superfluous quantities and entities from a theory as effectively as instance testing. One worry here is – setting apart the fact that the notion is slippery and has many meanings – that simplicity in a theory depends generally on the choice of the language it is cast in, and even on the choice of the theory's axioms. Language-dependence, however, does not seem to affect much the weighing of *comparative* simplicity in 'similar' theories. Theories like $T1$ and $T2$ share a basic language form and even most axioms.

What is needed, it seems, in order to eliminate this source of underdetermination is a principled reason to believe that hypotheses involving unobservable quantities or entities as part of a theory are actually false or likely to be *false*. Theoretical alternatives with superfluous extra baggage, which are in other respects observationally hd-indistinguishable from its more parsimonious theoretical rival, can then be dismissed with justification. Testing by instances does not provide such a reason. It caters to the 'prejudice' that leads scientists to prefer physical quantities in a theory to be all measurable; it does not explain or justify the preference. Accounts of hd-methodology score no better in this matter.[34]

I doubt that a principled, general reason for the methodological truism exists. For, future developments (the conjunction with novel theories) may turn a hitherto untestable, isolated hypothesis in a theory into a testable one, an unobservable redundant quantity into a measurable one. In addition, strict adherence to the "truistic" rule would have removed theoretically and empirically highly successful theories from the scientific body of knowledge. Bohr's first model of the atom, the "old quantum theory", correctly accounted – by straightforward deduction – for the line spectrum of hydrogen (the Balmer and Paschen

---

[33]Glymour may think otherwise, judging from a running example and his discussion of Reichenbachian underdetermination of space-time theories (Glymour, 1980b, pp. 357f).

[34]In some cases probabilistic reasoning may offer the reason sought after. For instance, since $T2$ entails $T1$ the probability of $T2$ is smaller or equal to the probability of $T1$.

series) and other ordinary elements. Part of Bohr's "planetary" model of atom was the assumption that the electrons have a well-defined velocity (frequency of revolution) on their stationary orbits around the nucleus. This theoretical quantity is indeterminable by measurements, since it is not related to the frequency of emitted light quanta. The lack of empirical determinability did in fact little to diminish the support the theory drew from verified deductions (Hermann Weyl was skeptical), and did not stop progressive elaboration of the theory.

In sum, there is little ground to think that instance confirmation is 'fit for the purpose'. This is not to deny that in concrete cases, testing by instances may shed light on, or even justify scientists' choices when faced with mutually underdetermined theories. Empirical strategies for determining values of parameters in a theory are an important issue in testing; predictions are often not possible without those parameter values in place.[35] But instance confirmation does little to block general arguments for strong underdetermination of theories.

3.4 Demonstrative Induction

The pursuit of this section is to assess the methodological significance of a variant of the method of eliminative induction: "demonstrative induction". This is essentially a mode of deductive inference in which a hypothesis is *deduced from the phenomena* and general background assumptions. Deductions of this kind are well established in scientific practice.[36] The deduction of a hypothesis from phenomena and unproblematic background knowledge apparently achieves two things that proponents of *underdetermination* and the *Duhem–Quine* thesis tend to deny: (i) in some cases at least exactly one theory is compatible

---

[35]The "standard" theory in particle physics, which is considered as a shining intellectual achievement, depends on 32 such parameters, including various particle masses.

[36]Its importance has been recognized by many a philosopher of science since the times of John Herschel, for instance, Dorling (1973), Norton (1993), Laymon (1994), Bonk (1997); it also goes under the names "quasi induction", "qualitative induction" and "Exhaustions Methode".

with the data, up to notional variants; and (ii) the allocation of blame for the failure of a rival theory can be located precisely within its system of hypotheses, regardless of auxiliary premises. In other words: under suitable circumstances the Duhem–Quine thesis is defeated, and so is hd-indistinguishability, insofar the former thesis supports the latter (Norton, 1993, 1994). I will argue that demonstrative induction does not invalidate either of the two theses.

W. E. Johnson, the Cambridge logician who made this mode of inference fully explicit, used as illustration *natural kind hypotheses*, like the deduction of the empirical regularity "All samples of argon have atomic weight 39.9." from the *major premise*: "All samples of any given element have the same atomic weight." and the *phenomenological premise*: "This sample of argon has atomic weight 39.9." The illustration makes it plain that demonstrative induction can be interpreted as a kind of Millian *eliminative induction*. Surprisingly, this kind of argument supported a number of important theoretical innovations, like Coulomb's and Cavendish's investigation of the electrostatic force-law, Darboux's and Bertrand's investigation into the uniqueness of universal gravitation, Einstein's arguments for the quantum hypothesis of light, and Planck's quantum hypothesis (see, for instance, Dorling, 1973). The implications of Planck's theory of black-body radiation, for instance, were sufficiently radical for physicists to research the possibility of an alternative classical explanation, in terms of classical particle mechanics and electrodynamics, for the radiation data Planck sought to explain. R. Fowler then proved that Planck's radiation law strictly *implied* the integral quantization of the linear harmonic oscillator's energy spectrum, within the framework of statistical mechanics. This result strongly suggests that there is no alternative classical explanation, which accounts for the relevant class of radiation data (Norton, 1993). The results appear to prove that hd-indistinguishable, conflicting theories cannot exist in this case. The body of data "determines" a theory uniquely, although relative to a wider theoretical framework. Since there is no reason to believe this phenomenon is unique to statistical mechanics, atoms and black-body radiation, demonstrative induction may well be the key to the solution of the problems a thesis of strong underdetermination poses.

3.4. Demonstrative Induction

I propose to illustrate the strengths and shortcomings of demonstrative induction in a somewhat simpler and accessible, but otherwise similarly situated case: the deduction of universal gravity in the context of Newtonian particle mechanics from observation on binary stars.[37]

The deduction of Newtonian gravity against the backdrop of classical mechanics has at its core a thought provoking mathematical fact. The formal physical setting is that of the non-relativistic Kepler problem in celestial mechanics. The difference to the traditional Kepler problem is that the force is not assumed to be universal gravity. Suppose the force in the so-called equivalent (one-dimensional) two-body problem is a central force; and assume the force acting on the particles is such that all bounded trajectories of a particle are re-entrant. It follows that exactly two types of attractive force laws are viable: either the force is proportional to the distance between the two bodies ("Hooke's law"), or the force law is inversely proportional to the square of the distance. This is the content of *Bertrand's theorem*, which asserts that the only forces compatible with the existence of re-entrant orbits for (almost all initial values of energy and torque) are linear forces in the radial distance, or the inverse square law of gravitation. The existence of universal gravitation can be deduced from Bertrand's theorem, provided the conditions of the theorem hold, either by an *experimentum crucis* between the two force laws (hardly necessary), or by imposing a boundary condition to the effect that the force vanishes for $r \to \infty$. Bertrand's theorem is a superior alternative to the (problematic and much discussed) method Newton, who started from Kepler's laws, used to "determine" the form of the law of universal gravitation. (It is mainly the theoretical innovation of the gravitational *potential*, which barred Newton and his contemporaries from discovering and exploiting the theorem, and only to a lesser degree the particular astronomical data which Bertrand had access to.)

Bertrand's argument is a paradigm of demonstrative induction. The (independent) evidence for The re-entrance hypothesis of orbits

---

[37] For additional details see Bonk (1997); an interesting case of demonstrative inductive reasoning in *quantum field theory* is presented in Bain (1999).

plays the role of the *phenomenal premise* in the argument. It is supported by excellent (independent) evidence: the planets revolve on ellipses and even binary stars (e.g. objects outside the solar system) trace closed orbits. Gravitational perturbations from bodies near and far account for any exceptions from the rule (the setting is non-relativistic). As a generalization from telescopic data the hypothesis is arguably not worse off than any low-level phenomenological law. The *major premise* is the conjunction of dynamical equations, the constraint on central forces and the stability assumption. The framework of Lagrange particle mechanics is part of the background theory. (These divisions may vary with the formulation of the argument.) The upshot is: instead of a bold conjecture about what force law governs nature, we are only asked to accept two relatively weak, independent assertions. The "facthood" of the re-entrance hypothesis and the relative generality of the dynamical setting make them resistant to criticism and force the rejection of a considerable segment of the boundless space of alternatives to the Newtonian gravitation hypothesis. The inverse square law of gravitation, deducible from fairly entrenched, logically weak and plausible premises, thus appears as "inevitable" and "near certain". Such is the force of demonstrative induction.

On closer analysis, the justificatory force of demonstrative induction is seen to depend on the satisfaction of four requirements. These are necessary and sufficient to justify the conclusion of a demonstrative induction. It is easy to see that these requirements are satisfied in Johnson's atomic weight hypothesis (with the exception perhaps of the last condition), and for hypotheses involving generalizations over natural kinds in general.

DI1 The major premise is better, independently confirmed than the argument's conclusion.

If the major premise would be less confirmed than the derived law, the argument would not carry any weight. With hindsight there is always a major premise to allow for the demonstrative induction of any given hypothesis from any data. Let *major* premise G be the disjunction of H and H'. H' is an arbitrary hypothesis for which the following relations are true: $\neg(H\&H')$, $H \to O_1$ and $H' \to O_2$. Verification of $\neg O_2$ by experiment may be taken as the minor premise, and the demonstrative induction argument supports or "selects" H accordingly. However, since H' is an arbitrary dummy hypothesis H has not become thereby "more certain". This indicates, perhaps, additional formal requirements of the major premise need to be imposed, besides it being sufficiently well confirmed.

DI2 Major and minor premise *retain* their initial degree of confirmation regardless of how improbable the conclusion of the argument may appear to be.

If the requirement were not satisfied, a demonstrative induction can be turned into a *reductio ad absurdum* for the major premise, say, whenever the conclusion, typically an intermediate-level law, clashes with other beliefs.

DI3 The minor, phenomenal premise states a relevant, specific, independent set of data, or else a (true) low level generalization thereof, such that if the data were different from what they are, the potential conclusion from the argument would be a different one.

The requirement is motivated by the need for a *contrasting set of data* such that the major premise together with the contrasting set would entail a different and with the previous one incompatible conclusion. For instance, DI3 is not satisfied in Fowler's derivation of the quantum hypothesis for the linear oscillator.

DI4 The minor, phenomenological premise is stated in an *autonomous* observation language.

The requirement states that the terms of the observation language are defined, or fully interpreted, in particular without involving the theory to which the major premise belongs in any form. The concept of wavelength (of electro-magnetic radiation), for instance, for use in the phenomenological premise is supposed to have an operational definition. We want to be able to use the term without committing ourself to one or the other idea of what, for example light is according to one theory or another.[38]

These four requirements are not satisfied in many intended reconstructions of historical cases, which display reasoning by demonstrative induction. Among them is Bertrand's derivation of universal gravitation, which is structurally similar to arguments advanced by Cavendish and Coulomb for the inverse square law of electrostatics as reconstructed by Dorling.

Few will, after reflection, seriously consider the demonstrative induction argument for the inverse square law outlined above as justifying belief in it. First, the major premise in Bertrand's derivation, the dynamical equation for the general two-body problem without specification of a force law, is difficult to test and to confirm. It is also *less* falsifiable than the inverse square law itself. For any test presumably passes through the specification of a model, and blame for a falsified prognosis will be allocated to the force law, or any other model assumption, not to the general dynamical scheme. Another method to confirm the dynamical equation directly is the elimination method. But scientists' confidence in Newton's three laws (and the Lagrange formulation of particle mechanics) did not spring from a comparison with other systems of fundamental laws, nor would an exhaustive comparison be feasible.[39] Consequently there is little reason to believe the major premise is more certain or better confirmed or tested than the consequence: universal gravitation.

Secondly, the other assertion of the major premise, the "central force" constraint in Bertrand's theorem (not taking into account the extra requirement of a stable circular orbit), excludes a priori velocity- and direction-dependent potentials. At the same time the constraint permits as "possible" alternative potentials *a priori* implausible one like $kr^{-5.5}$ (the solar system as we know it would not exist by now). This is clearly not a well-chosen reference class. Its selection is *ad-hoc*:

---

[38] Like the foregoing requirement this one was prompted by the examination of the significance of Fowler's result, see Bonk (1997). Paul Feyerabend has argued at length that the 'detachment' requirement is implausible.

[39] Demonstrative induction appears to presuppose a more basic account of testing and confirmation, like hd-methodology, in order to answer these difficulties.

independent of the empirical success of Newtonian gravity there is little reason to single out central forces. The difficulty here goes beyond delimitating a class of alternatives in the right manner; but also of inventing *strong* rival alternative hypotheses.

Finally, consider a hypothesis for the gravitation potential that is for practical purposes *hd-indistinguishable* from the true one, say a powerlaw in the radial distance like $r^{-1.0000001}$. *If* one accepts this law (along with the dynamical equation and the central force assumption) one will have to reject the "re-entrance" hypothesis of Bertrand's theorem, i.e. the minor phenomenological premise, to avoid inconsistency. The phenomenological premise was taken to be fact-like. Yet, this is an idealization and the premise is a low-level generalization based on a finite number of observations and a margin of error. Rejection is reasonable once doubt is sufficiently motivated. A dilemma looms here: if one believes in Newtonian gravitation the phenomenological premise follows and appears as confirmed. If one does believe in the alternative potential (on independent grounds) the premise does not follow and one has all the reasons to reject the re-entrance claim in the minor premise. So the justification of the minor premise presupposes the conclusion of the argument. The independence condition of DI3 is not satisfied.[40] In conclusion, Bertrand's theorem cannot justify belief in universal gravitation, by eliminating rivals to the gravitation hypothesis, and is not in any way a superior route to gravity than a bold Popperian conjecture. The "inevitability" of the Newtonian force law is bought by artificially restricting the space of competitors, and by taking for certain what is far from certain, whose certainty depends on what is to be proved.

The objection to justification by demonstrative induction, illustrated by the example above, is not that the antecedents of the argument are not above doubt or that they rest themselves on inductive inferences. Rather one or more among them is (a) less testable and less tested than the argument's conclusion, and (b) that the certainty or

---

[40]Note the *explanatory asymmetry*: while the conclusion, universal gravitation, *explains* the fact that the orbits are closed within the framework of the major premise on one account of explanation, it is not the other way around.

testability of the antecedents are dependent on the certainty or testability of the conclusion. This is not to deny that judgements about the uniqueness of a theory or hypothesis in scientific practice often follow a pattern marked out by demonstrative induction. The pattern seems also ideally suited to theory- or framework internal determinations of parameters in a model or theory. And it is one explanation for why issues of underdetermination are not typically foremost in the mind of scientists.

3.5 Underdetermination and Inter-theory Relations

In the present section I address the claim that a theory can gain indirect confirmation through the relationships it bears to other theories. The claim, if true, pinpoints a potential shortcoming of the hypothetico-deductive methodology, and shows that empirical hd-equivalence between rival theories may be of little significance.

This possibility is explored in a series of papers by Larry Laudan and J. Leplin and it is to their ideas that I turn first. Can a verified observational consequence *fail* to confirm the theory from which it has been deduced? Can a theory pick up confirmation from verified observation sentences which are *not* consequences, i.e. which do not belong to the theory's observational hd-content? In the first section above I have recognized reasons for an affirmative answer to the first question. I now turn to the second question.

> Being an empirical consequence of a theory is neither necessary nor sufficient to qualify a statement as providing evidential support for the theory. Because of this, it is illegitimate to infer from the empirical equivalence of theories that they fare equally in the face of any possible or conceivable evidence. The thesis of underdetermination [...] stands refuted.[41]

---

[41] Laudan (1996, p. 68). The paper is jointly authored by J. Leplin; compare Leplin (1997a, pp. 154, 158).

The claim is that the set of evidentially relevant data for a deterministic theory overlaps with its deductive hd-content, but does not in general coincide with it. Darwin's theory of selection has been advanced as an example, since it had limited hd-content and allowed few predictions, yet surely is well confirmed (Miller, 1987, p. 244). The Darwinian theory explains and unifies a lot of diverse facts, and the latter make the theory 'probable', not to mention other factors that contribute to its towering observational support. Yet, the move to probabilistic or explanatory considerations for or against a theory implies the adoption of a theory of confirmation quite different from plain hd-methodology. This move automatically brings with it recognition of a broader set of evidentially relevant facts than the hd-account admits. A theory's observational content varies with many factors, one of them is what one considers as observational support. The example thus does not weigh in decisively against the coincidence thesis of evidence and content.

Inter-theoretical *analogies* may have evidential relevance. One of two hd-equivalent theories, but not the other, may be 'analogous' to a highly successful theory in another domain. Or, one of two hd-equivalent theories, but not the other, is entailed by a highly successful theory of greater scope and generality (Laudan, 1996, p. 67). To begin with the latter possibility,[42] consider the case of two theories T_1 and T_2, which are conceptually distinct yet deductively hd-equivalent (have the same hd-content). Assume that the first is derivable from a more comprehensive theory, T, while the second is not. Assume further that the comprehensive theory is independently hd-confirmed by a pair of observation sentences O_1, O_2, representing initial datum and entailed actual observation sentence respectively, and that the deductions do not depend on premises that encompass any hypotheses shared by T_1. (I suppress the dependence on O_1 from now on.) Laudan and Leplin claim that since O_2 supports T_1 indirectly *by way of* T, but not T_2, the two *are* inferentially distinguishable despite sharing the same deductive

---

[42]The conjunction of one theory with another one may lead to new explanations and novel predictions extending the support for one or both of the two. Configurations of theories like these reflect more, I think, on the openness of a theory's hd-content than on the falseness of the coincidence thesis of evidence and content.

observational content.

This is a first example of *inter-theory* relations adding to or subtracting from the empirical support of a given theory. The account above appears to entail that generally empirical support or confirmation is 'transferred' between sub-theories of a theory (like T). I doubt that this the case. Herschel's observations on the orbits of double-stars did nothing to confirm Kepler's laws, or *vice versa*, both sub-theories of Newtonian mechanics cum universal gravitation. Nor did the successful quantum mechanical account of the hydrogen spectrum incrementally confirm the quantum mechanical account of Schrödinger's cat, or *vice versa*.[43] Putting this worry aside, consider the hypothetical transfer of the confirmational impact of observing the datum expressed by O on T_1 in abstraction. Intuitively, in accordance with the *converse consequence condition*, the confirmation that accrues from the verification of the observation sentence O is shared "upwards" with all premises of the deduction. The second intuition behind the argument is, I must assume, that the empirical support of T is shared "downwards" with whatever sentences T entails, i.e. with T_1. Thus T_1, but not hd-equivalent T_2, is additionally indirectly confirmed. Now, the operating principle here is Hempel's *special consequence condition*, a principle of questionable methodological status. Moreover, both of Hempel's 'conditions' cannot be satisfied together, for their joint use implies that every sentence supports every other sentence. So the idea of a 'transfer of confirmation from T to T_1' is problematic.[44] There is a problem of consistency for the argument lurking here: if one assumes hd-equivalence of T_1 and T_2, then the confirmation relation should be be hypothetico-deductive in character too. And the special consequence condition is

---

[43]Examples like these lend credence to the belief that only data, which are *explained* by the theory in question, can add to its confirmation. A requirement like this can easily supplement the account of hd-confirmation sketched above (were it not for well-known uncertainties with regard to the very notion of 'explanation').

[44]This objection was first presented in Okasha (1997). Laudan and Leplin anticipate this criticism, but their *defense* of the special consequence condition rests on the misconception that the difficulties mentioned are generic to a narrow set of theories of confirmation, which take satisfaction by instances as their central notion.

not generally valid in a strict hd-account of confirmation. Yet, if observational equivalence is understood in a more liberal sense, including all 'relevant confirming data' whether in or out of a theory's deductive content, Laudan and Leplin's objection does not contradict the claim '$EE \to UD$'.

Laudan and Leplin stress that their example of how inter-theory relations may resolve empirical underdetermination amongst theories is only an example. The authors in addition point to "sophisticated analogical reasoning" of scientists as further evidence that evidential support can accrue from outside a theory's hd-empirical content (Laudan, 1996, p. 67). This interesting suggestion is made in way that is so brief and sketchy that criticism has to await a more full and detailed account. In the meantime, my guess is that it will be extremely difficult to entangle genuine evidential analogies from the heuristic or rhetorical use of analogies. In sum, I conclude that the arguments discussed do not present a strong reason to believe that "the thesis of underdetermination [...] stands refuted".

Richard Boyd's "causal theory of evidence" addresses the powerful intuition that if one but not the other of two observationally hd-indistinguishable rival theories shares significant theoretical characteristics with highly successful background theories, then it is more likely to be true. Again, a theory can accrue confirmation through inter-theoretical relations, i.e. through observation sentences not belonging to the theory's empirical hd-content, this time relations to background beliefs shared by scientists. Boyd developed this intuition by considering Poincaré and Reichenbach's (among others') contention that the geometry of space and time, and the laws governing particle dynamics in space and time, are tested together *as a whole*, which offers no natural partition in point of confirmation. So a little stage-setting is in order (compare section 2.6).

Denote summarily by F the physical theory in question and by G the underlying orthodox geometric assumptions. Admitting an undetectable 'universal' force to F which affects (deforms) the measuring instruments, measurements of distances, lengths, areas and angles will result in a different geometry, G' say. Specifying either directly the standard of length, or the form of universal force (and hence the rigidity of measuring rods) amounts to giving a 'coordinative definition'. Through mutually *compensating* adjustments the resulting cluster of laws $F' + G'$ becomes observationally indistinguishable from the orthodox $F+G$, if basic observation sentences refer to particle positions and times. Although these are hd-equivalent descriptions, according to Boyd, it would *not be rational* to accept the non-standard theory compound: the two are not 'methodologically' equivalent. There are significant structural similarities between $F + G$ and other independently successful theories, but there is no successful theory, which postulates a theoretical object similar to Reichenbach's 'universal' force. Since from the scientific realist point of view the empirical success of standard force-based theories is explained by the approximate (correspondence or "correlation" concept of) truth of these theories, the structurally similar $F + G$ is more likely to be descriptively true than the non-standard compound of laws. Hence, the rejection of $F' + G'$ is *rationally warranted*. It is worthwhile to quote the relevant passage at length.

> Furthermore, this estimate of the implausibility of "F' and G'" reflects *experimental* evidence against "F' and G'," even though this theory has no falsified observational consequences. This is so because the experimental evidence which led to the adoption of our current theories of force is evidence that there really are electrical, gravitational, magnetic, and other such forces *and* that they *all* do result from such matter-dependent fields as have been described. But, then, this fact [...] is, in turn, evidence that *all* forces have such an origin. So the experimental evidence for our current theories of force is indirect experimental evidence that no such force as f' [a "universal" force – T.B.] exists – and that "F' and G'" must be false. (Boyd, 1973, p. 8; emphases R. Boyd)

More generally, it is rational to believe the one among underdetermined alternatives, which is structurally most similar to successful, mature 'background' theories. The data, which support the latter become relevant and indirectly support one (or more?) of the hd-equivalent alternative theories. Therefore the set of confirmationally relevant data, again, exceeds a theory's deductive content. Conversely, it is reasonable to reject theories as likely false that postulate entities (forces, etc.), which have no counterpart in well-tested background theories, provided there is an hd-equivalent alternative whose entities do have such counterparts:

Background Rule From a class of underdetermined alternative theories it is rational to believe the one that is most compatible with our background information.[45]

Boyd justifies the rule by appeal to a meta-criterion: a methodological rule in the sciences is warranted if its employment *reliably* raises the likelihood that theories, which are accepted on account of the rule, are instrumentally successful. The reliability of the rule in turn is "best explained" by the approximate truth of the background theories in question.

Scientists have constructed theories or hypotheses with an eye to what has 'worked' in similar circumstances before. Questionable is not the Background Rule as an encapsulation of *research heuristic* in theory formation, but as a statement about a source of additional evidence for a theory. For one, it is ambiguous and so cannot effectively guide rational choice. Suppose the relevant successful background theory postulates entity X. Is the scientist required to believe only the one underdetermined alternative from a set of hd-equivalent theories, which postulates X as well, even if it does so at the *cost* of great complexity in its laws? Or, does the evidence that supports the background theory support another alternative from the same set, which postulates entity Y instead of X, where Y may or may not be similar in some respects to X, and its laws are of great elegance? These are not speculative possibilities. Heisenberg and Born were right to copy the formal structure of classical mechanics in the construction of matrix mechanics. Yet, they would have erred fundamentally if they had attempted to take over the otherwise so successful idea of particle trajectories. The Background Rule, as it stands, is too ambiguous as to

---

[45] My formulation of the rule is extracted from Boyd (1973, p. 9).

accurately state conditions for when a piece of evidence relates to one but not another underdetermined alternative in a set of hd-equivalent theories.

How reliable is the proposed rule in selecting the true theory? It is difficult to gauge its success, for there are too few examples of observationally underdetermined pairs of theories for this purpose. The historical record of the success-rate of inter-theoretic plausibility arguments for new hypotheses appears to be mixed (Miller, 1987, pp. 371, 398, 458). The rule would presumably have led to the rejection of the Copernican theory in favor of one of the versions of Ptolemaic theory, which was in line with accepted, successful background theory. Similarly, if one takes the Background Rule at face value the Bohmian theory of quantum mechanics, which retains the previously so successful idea of definite particle trajectories, is much better confirmed than its rival from Copenhagen. It is thus at variance with actual judgements of the majority of physicists. In addition, a mock rule like "from among a class of strongly hd-indistinguishable alternatives select the one which came historically first" (after L. Laudan) appears arguably to be as reliable as the Background Rule: Matrix mechanics and the orthodox Copenhagen theory were there long before Bohm's or Popper's accounts of micro-physics.

I leave Boyd's argument at this point with the remark that accounts of inter-theoretical relevance of data need more elaboration before they become effective against the (naive) identification of evidence for a theory with the hd-content of a theory.[46] In any case, the identification is not an essential part of theses of strong underdetermination.

Richard Miller too advances inter-theoretical relations as the key to the refutation of underdetermination theses. Miller's thesis is that underdetermination is *unique to physics* and basically a historical phenomenon of the past. The first part of the thesis follows from the perceived asymmetry in the 'frequency' of hd-equivalent theories in

---

[46]See Douven (1996), who focuses on the part the 'miracle argument' for scientific realism and the inference to the 'best explanation' plays in the argument, concerns I share, and Miller (1987). Does the 'pessimistic meta-induction' from the eventual failure of past theories cast further doubt on the supposed approximate truth of our current background information?

3.5. Underdetermination and Inter-theory Relations

physics and in the other natural sciences; I have discussed the argument in section 1.4. The second part follows from the general premise that natural science in essence is a search for *causes* (Miller, 1987, pp. 441, 177). Now, the history of physics differs from the history of chemistry or botany, say, in that it was characterized for a long time by a huge and persistent *gap* between the knowledge of phenomenological regularities and knowledge of the physical causes of these regularities (Miller, 1987, pp. 431f). Physical theories were instrumentally successful to the extent they were, yet insight into the causal processes explaining the phenomena was lacking behind.[47] As the physical sciences progress causal knowledge accumulates and a unifying description of natural processes develops. Each novel, successful theoretical advance in the history of physics helped closing the gap. Scientific progress tends to *eliminate* precisely those hypothetical objects and quantities from our theoretical description of the phenomena, which give to underdetermined alternatives. For instance, to rehearse a well-known example, in Newton's cosmos the absolute velocity of the center of the universe is underdetermined, and alternative theories may disagree over its true value and direction. Relativity theory shows how the very idea of absolute space and time, along with the concept of an ether and absolute motion, can be *eliminated*, and with them a *source* of underdetermination. Similarly, Miller maintains that since general relativity theory "geometrizes" the Newtonian gravitational potential away Reichenbach's examples are obsolete.

> But its rejection [the "non-standard catalogue" of forces in $F' + G'$] is justified by Einstein's basic principle of invariance, and by the (hedged, prima facie) principle that action-at-a-distance does not take place, when these are combined with data about electromagnetic phenomena. The result is relativity theory, in light of which the uniform distorting force hypothesis can be disproved. The ultimate background principles, here, are not preferred as a mere matter

---

[47] Today, the plurality of theoretical models in nuclear physics illustrates the explanatory gap, according to Miller.

of convention or of antiquity. Rather, they are supported by data combined with truisms. (Miller, 1987, p. 433)

This interesting argument rests on the diagnosis that *all cases* of underdetermination spring from the postulation of physical entities or magnitudes, which the theory in question somehow does not allow to determine or to test by observation (compare section 3.3). Theoretical structures like these do engender hd-equivalent alternatives. However, arguments for a strong underdetermination thesis may take off from other premises, for instance, the Duhem–Quine thesis or sheer confirmational indeterminacy in hd-confirmation, conventionalist moves, algorithms, or in yet other ways. Miller's argument cannot be considered effective as long as these by-ways to underdetermination are not blocked.[48]

Suppose Miller were right in identifying the source for the underdetermination of theories, or suppose other arguments for underdetermination fail. Suppose furthermore that the history of physics is such that new advances in theory eliminate those sources in predecessor theories. What follows for the truth of our thesis? The thesis does not stand refuted, for the latter is *not a diachronic* claim, but a synchronic one: at every time, any theory (or total science?) of sufficient power has observationally equivalent alternatives. He has not shown that current theories do not have the 'defects' (with respect to underdetermination) of some of the predecessors they replaced. For instance, consider the claim "The dramatic reduction of equivalence in the recent past is the triumph of relativity theory" (Miller, 1987, p. 436). Perhaps, but the advent of general relativity theory has not rid us of "universal" forces.[49] Reichenbach wrote: "whereas Einstein's definition of ds^2 [the infinitesimal line element squared – T.B.] in terms of clocks

---

[48]I find little on alternative arguments for systematic underdetermination in Miller (1987).

[49]Miller dismisses, contrary to Boyd and others, the Reichenbachian pair as a flawed example of the thesis of underdetermination that fails to support it. The alternatives $F' + G'$ share too many physical objects with the orthodox theory (entities and standard forces) and hence are not a sufficiently radical departure from the orthodox theory to count as "genuine rivals" (Miller, 1987, p. 432).

3.5. Underdetermination and Inter-theory Relations

and rods excludes universal forces from the *four-dimensional* manifold, such forces now are unavoidable in *three-dimensional* space. [...] the transforming away of universal forces is no longer completely in our hands" (Reichenbach, 1958, p. 263; emphases H. Reichenbach). In any case, there is nothing in current gravitation theory that prohibits topologically alternative space-times some of whose events are causally inaccessible Malament (1977), or theories like Thirring's (Thirring, 1961). Notoriously, non-relativistic micro-physics is accounted for by hd-indistinguishable theories, which in addition cast some doubt on the utility of the notion of cause.[50] The upshot is that scientists witness at least as much underdetermination in the most current theories than their precursors in theirs.

I have not been persuaded by the claim that science is the search for causes. I find it difficult to distinguish on the basis of the historical record as our datum between Miller's interpretation of history and 'science is the search for explanation' or 'science is the search for (approximate) truth' for instance. These interpretations, or meta-narratives, all fit the record of scientific progress more or (often) less well; but the latter two do not necessarily invoke 'causes'. I tentatively conclude that the "history" of underdetermination in physics, as sketched by Miller, does little to shake confidence in strong underdetermination.

In the final paragraphs of this section I like to address another promising proposal on how to eliminate observationally indeterminate physical entities or magnitudes, which engender strong underdetermination (compare section 3.3). The methodological principle Michael Friedman suggests centers on the notion of *unifying power* of a theory. A unified theory is more than a conjunction of independent phenomenological regularities, or models. A simple schematic example is the kinetic theory of gases, which unified various partial and local laws, like the well-known relation between pressure, volume and temperature in an ideal gas (Friedman, 1983, pp. 238f). What this 'more' amounts to is difficult to say in the abstract, although some progress has been

---

[50] A realist version of quantum mechanics, roughly along Popper's propensity theory, is defended in chapter 11 of Miller (1987). I remain skeptical, but assessing Miller's interpretation is beyond the scope of this essay.

made. All agree that the gain in unifying a set of independent laws into one tightly knit structure is confirmational: in a unified theory, evidence for a sub-theory can become relevant evidence for an otherwise unrelated sub-theory. A theoretical aspect (entity, force, state of a system) that increases the internal "unifying power" of a theory increases potentially the theory's overall empirical support (and *vice versa*?).[51] Fortunately, it is possible to assess the utility of the principle with respect to our present purpose without having a formal and fully acceptable account at hand. For Friedman is concerned with the problem of restricting the theoretical apparatus of a theory, as much as possible, without going all the way to the theory's Craig sentence and eliminating theoretical expression altogether (Friedman, 1983, pp. 28f). To permit no more theoretical superstructure (for a domain of experience) than a *unified* theory for the domain requires, is *one way* to draw a line between the two extremes: arbitrarily de-Ockhamized superstructures or a fully Ockhamized infinite set of observation sentences.

How does this approach impact on the issue of hd-equivalent theories? Consider a special sort of example, namely a theory, which admits an object or quantity that cannot be 'determined' by measurement and theory together; its observational equivalents make contradicting assignments of a value of that quantity, compensated by suitable changes in other places of the theoretical structure. (The result is a thorough systematic underdetermination.) In the well-worn Reichenbachian case, for instance, the universal forcefield in $F' + G'$ *lacks unifying power* since the universal force F can be replaced indiscriminately by different expressions $F', F'', \ldots$ and hence does obviously not contribute to the theory's potential for unification (Friedman, 1983, pp. 299–301). Generalizing, one is led to the following rule:

Unification Rule From a class of underdetermined alternative theories reject all those that postulate theoretical objects "that are indeterminate and have no 'unifying power' in a given

---

[51] Friedman's account does not appear to run afoul of the special consequence condition, see Friedman (1983, p. 244, n. 15).

3.5. Underdetermination and Inter-theory Relations 137

theoretical context".[52]

The implicit assumption is, of course, that the class contains at least one and not too many theories, which do *not* postulate theoretical objects "that are indeterminate and have no 'unifying power'". Reichenbach's example shows that this condition for the applicability of the Unification Rule is sometimes satisfied.

Two reasons can justify such a rule. One reason is an argument to the effect that a unified theory is better confirmed than any of its non-unified hd-equivalent rivals by data from the same observational content. Second, a good reason to adopt the rule is its proven or likely reliability in picking the true, or most likely theory from the set of its hd-equivalents. To begin with the last possibility: the chances to show the reliability of the rule are slim. How would we find out that an alternative not de-selected by the rule is true or likely to be true (provided there is only one such theory left after parsing)? Since all theories in the (very hypothetical) set of its hd-equivalents stand and fall together in a test, deductively won data give no indication. A diachronic perspective may advance the case: possibly only one theory in the set shares significant theoretical characteristics with its *successor* theory. This is a possible retrospective indication that the precursor was 'close to the truth'. Similarly, one and only theory in the set may prove to be fruitful in conjunction with other theories. This fact too may count as an indication that the theory is 'close to the truth'. We would then check if this theory is unified or not. If the two properties coincide frequently enough the Unification Rule may count as reliable. Unfortunately this is, obviously, a rather speculative procedure; there are far to few historical data to test the rule against. Therefore the reliability argument in favor of the rule is unlikely to succeed.

Is a unified theory better confirmed than any of its non-unified hd-equivalent rivals by data from the same observational content? If 'confirms' here means 'hd-confirms' then the answer, based on the plainest form of hd-methodology, is no. If 'confirms' means 'probabilistically confirms' then the answer, based on a Bayesian account, is maybe

---

[52]I do not know whether Friedman would endorse the rule, but compare Friedman (1983, p. 294) where the quote is taken from (p. 338 and p. 29 ibid.).

yes. This yes, however, essentially reflects a suitable choice of prior probabilities in favor of the unified theory in the set of hd-equivalents. Confirmation by satisfaction might best fit the bill. But not only has it difficulties by itself (see above), confirmation by satisfaction or instances would make appeal to the criterion of unification redundant.[53]

Judgements about the confirmational status of a unified theory cannot draw on the intuitions that made the introduction of the concept of unification plausible in the first place. The eventual replacement of an independent set of phenomenological laws by a 'unified' theory may well increase the observational support of each of the phenomenological laws, taken now as consequences of the unified theory (waiving the difficulty that the consequences are typically not identical to the original laws). It is far less clear how to account for the boost of confirmation the unifying theoretical superstructure itself receives, if it does receive a boost at all. In any case, the unified theory will have excess hd-content and may be confirmed by novel predictions, etc. The problem that interests us presently is essentially different in that unified and non-unified alternative have exactly the same hd-content.

I conclude from these brief considerations that the elimination criterion of unification, like a maxim of simplicity, needs back-up from a detailed account of confirmation, before it can figure in a satisfying reply to one argument for systematic strong underdetermination. No suitable candidate is at hand at the moment. Let me take stock. One wants to say that all but one of a class of observationally indistinguishable, rival theories depend on "bizarre" or "superfluous" or less-confirmed postulates or indeterminate quantities. I cannot rule out that one or another methodology-based rule can settle the underdetermination issue in general. But the ones I have extracted from the work of Friedman and Boyd are not the right rules, for they have not been shown to be truth conducive. There is no knockdown argument to be discovered here. The best I can do is to diffuse the intuitive support these rules enjoy. I accept these rules as rules of thumb – with

---

[53]Friedman gives not much indication of what he takes the confirmation relation between theory and datum to be; but a Bayesian account and plain hd-methodology can be ruled out (compare Friedman, 1983, pp. 241f).

considerable heuristic value – awaiting a proper grounding based on a detailed theory of confirmation.

Chapter 4
Competing Truths

I examine "constructivist" arguments for the truth of the thesis of strong observational underdetermination of theories or 'global' or 'total' theories, ranging from attempts at pythagorizing theories, to the constructibility of alternative language-forms and the generation of alternative theories by 'algorithms'. The arguments, if correct, tend to underwrite claims to universality and necessity of observational underdetermination or empirical irresolubility.

4.1 Constructivism

Historically speaking, "phenomenological constructivism", represented by Carnap, Goodman, Ayer and to a lesser degree by Reichenbach, reflected the conviction that any empirical content can be re-packaged in various conceptual frameworks, which differ in their logical scaffold and their choice of primitives. Discourse about tables, apples and clouds can, it was believed, ideally be construed or re-constructed as indirect talk about convoluted classes of "sense-data" or "electron-moments", percepts and the like. There "is a great number of possible system forms from which to choose" asserts Carnap (Carnap, 1963c, p. 946). I briefly survey in this section the variety of systems proposed with emphasis on the formal possibilities of generating alternative systems in systematic fashion.

Carnap's "Konstitutionstheorie", based on a few basic, non-logical relations operating on a simple universe of discourse, pioneered the constructive approach to knowledge of the external world. In *Der logische Aufbau der Welt* Carnap chose subjective *Elementarerlebnisse* – global and integral appearance-events – as the base set. This choice allows for making assertions about the sense content of any given place ("Stelle") of the otherwise non-analyzable whole of experience. The only non-logical primitive relation of the *Konstitutionssystem* is a dyadic (asymmetric) similarity relation ("Ähnlichkeitserinnerung") between pairs of *Elementarerlebnisse*. The method of construction in the *Aufbau* is explicit definition of new concepts within the requirements of strict extensionalism. The standard of reduction to the phenomenological basis is logical reduction or translatability of all descriptive common and scientific discourse (Carnap, 1961, pp. 46, 1). Reconstructing or reducing a given common-sense or scientific concept $P(x)$ thus amounts to replacing it with an extensionally equivalent combination of different basic predicates in all propositions in which P occurs in such a way that the agreement in extension is lawlike and non-accidental (Carnap, 1961, p. 46).[1] Carnap's paradigmatic examples for his procedure are the Frege-Russell reduction of Peano arithmetic and the reduction of the real number system to the system of natural numbers. On the solipsistic basis of *Elementarerlebnisse* objects of greater degree of complexity and intersubjectivity can be defined in successive steps (if only schematically and programmatically): regions in the visual field, colors, temporal and spatial order, visual presentations of things ("Sehdinge"), the body, physical things, other humans and their states of consciousness.

In §154 of *Der Logische Aufbau der Welt* Carnap remarked on the definability of a *new* and alternative set of primitive relations for the "Konstitutionstheorie", generated and induced by permutations among the basis elements. He felt the need to single out "erlebbare, natürliche" primitive relations as against the more artificial choices, and introduced to this effect a primitive, indefinable second-order relation: "fund", in a somewhat *ad-hoc* fashion. Besides subjective "Elementarerlebnisse" in the stream of consciousness Carnap considered alternative bases for the logical framework with *physical* elements (Carnap, 1961, para. 62). For instance, a finite or countable infinite set of electrons, structured by spatial and temporal relations, serves as

---

[1] The additional clause was urged by Goodman and accepted by Carnap in 1961.

the basis of one *physicalistic* framework. The basic spatial and temporal relations allow, according to Carnap, to define the particles' accelerations and the definition of fields and forces through relative accelerations. Statements about the action of gravitation can be reduced to (translated into) statements about relative accelerations of atoms and molecules. *Events* provide another alternative *physical* basis for the constitution of the external world: the point-set of the four-dimensional continuum (structured by topological relations and tuples of numbers associated with every individual, corresponding to field strengths, etc.), or segments (moments) of particle world-lines in Minkowski space. This sketch of Carnap's project already shows that key elements in a systematic account of our knowledge of the world, the individuals and primitive predicates, are "negotiable". Alternatives frameworks or 're-constructions' conflict with respect to basic, extra-logical assumptions and the choice of the universe of discourse. For example, the cardinality of the basis of a strict phenomenalistic, "nominalistic" system is *finite* (Goodman, 1966, p. 141). Different choices result in variant systems which, for Carnap, agree in just two points: in (ideally) saving the same phenomena and in using a *Principia* style, type-theoretic logical system. In Carnap's (later) view the various possible systematizations cannot conflict in a way that could give rise to sceptical concerns. For instance, the claim "sense-data exist" is not a factual claim about what there is in the world, according to Carnap. Rather it reflects a framework decision to use one particular from among a set of alternative frameworks, subject only to considerations of an expedient conceptual economy. Conflicting statements about what exists are always reducible to conflicting decisions about semantical and syntactical *conventions*. If defendable, this is an elegant and plausible way to both agree with and *substantiate the thesis* of strong underdetermination, and avoid its sceptical consequences. The difficulties 'framework conventionalism' faces are discussed in chapter 5.

N. Goodman's phenomenological system improves on various technical features of the *Aufbau* (Goodman, 1966). Here as there the basis is *psychological*, the qualia of place, of shades of color and of time. Place refers to place in the visual field, time to the temporal order between "appearance-events". The primitive dyadic relation in the mereological calculus of individuals is "togetherness" between the three types of qualia. Order among qualia is imposed by the dyadic, non-transitive "matching"-relation M. Carnap's requirement for the proper reconstruction of a predicate, identity of extensions, is replaced by the weaker global requirement of "extensional isomorphism", or identity of structures of the reduced and the reducing system. This framework suffices to constitute particulars with qualities like shape and size, and erect the scaffold for the description and structure of appearances. Characteristically, Goodman makes no claim to the epistemological primacy of the construction basis. Since he rejected the framework *conventionalism* of Carnap's "Empiricism, Semantics, and Ontology", Goodman interpreted the multitude of constructive systems as a multitude of *worlds*, each with its own universe of discourse. Because each system is "right" (in the sense of constructed in accordance with a few formal meta-requirements), and since the systems cannot be right in *the same world*, the systems are each right in their own world. This kind of strong ontological relativism, however, is as we shall see by no means an inevitable consequence of the observational indistinguishability of the systems.

Finally, A. J. Ayer outlined a "construction of the physical world" with a base higher up in the hierarchy of constructible things. The atomic elements comprise qualia of color, size and shape, as well as recognizable "visual patterns", a visual chair-pattern, a cat-pattern and so on. The primitive relations between qualia and patterns are spatial and temporal relations (simultaneity, precedence). In contrast to Carnap's *Aufbau* Ayer did not claim that sentences about ordinary things can be reduced, by chains of explicit definitions, to sentences about qualia or visual patterns. Rather, a final *posit* of (kinds of) ordinary things is required to complete the passage from universe of qualia to the external public world. Appeal to the best possible explanation, according to Ayer, makes the posit "probable", given the general structure of appearances (Ayer, 1982, p. 108).

The main objection against phenomenological constructivist programs, from the *Konstitutionstheorie* to Goodman's systems or Ayer's, is that none of the proposed systematizations, on whatever chosen basis, has achieved its goal: a phenomenalistic account of the physical world. All, including the physicalistic systematizations remain programmatic and sketchy to different degrees. None is, due to limited and fragmented content, a genuine alternative to natural and scientific language, however unsystematic, inconsistent and lacking in formalization these are. Quine, in a review of Goodman's *Ways of Worldmaking* took the "lack of full coverage" as the decisive objection against the proliferation of Goodmanian worlds. There are good reasons to believe that this historical failure of the constructivist program to provide a universal framework, from which eventually alternative frameworks can be systematically generated, is not accidental. Consider for instance the discrepancies between *physical time* and phenomenological time ("time qualia"). Statements and comparisons of physical duration would have to be defined by the notion of fluctuating and finite phenomenological duration (in the lifetime experience of the subject). Goodman wrote that the problem of physical time is "not easy" and lies outside the scope of his book (Goodman, 1966, pp. 377–378); but nor has the more general problem of "constructing [...] the physical object" out of presentations be solved (Goodman, 1966, p. 129). He has deflected the requirement of *compleatebility* of the program by way of an analogy: we do not reject Euclidean geometry because "angles cannot be trisected with straight-edge and compass" (Goodman, 1963, p. 551). True enough, however, we would have actually reject Euclidean geometry if it would not have allowed derivation of the Pythagorean theorem and other geometric truths it was intended to axiomatize.

The semantical difficulties for carrying out a global reduction or translation of thing language sentences, say, to sets of sentences in the experiential language point, perhaps, to fundamental obstacles for any such program. A sentence that expresses a simple observable state of affair in the material object language would have to have a translation into a conjunction of an *infinity* of ill-assorted atomic sentences concerning private impressions. It is impossible, however, (given the means of a natural language and cognition) to state the truth conditions for the former in a sense-data language (Reichenbach, 1938, pp. 101–102; Ayer, 1982, pp. 106–107). Another difficulty for the program is the 'resolution' of contrary-to-fact conditionals. Everyday discourse relies on causal hypotheses regarding the physical behavior of objects, as well as on natural laws. The explication of causal hypotheses and natural laws requires the acceptance of certain contrary-to-fact conditionals. These conditional sentences would need to be translated or reduced as well, and it is difficult to see, and never has been sufficiently explained, how this translation can be effected in terms of sequences of *actual* raw experiences (Ayer, 1982, pp. 151–152). Goodman, though, developed an account of counterfactuals that made progress toward resolving this difficulty in *Fact, Fiction, and Forecast* – an account that is, albeit, controversial.

One of the main reasons for Carnap to give up the "Konstitutionstheorie" is the objectivity problem: bridging the gap between the solipsistic construction based on private sense-data (or qualia) and publicly accessible facts. It appears that any explanation of intersubjective agreement between observers would require those observers to perceive the same sequences of sensations. Earlier on Carnap had taken the position that the choice of the subjective, "solipsistic" basis is a methodological, technical convenience for the purpose of system building. This position tacitly presupposes that one *could have* used a different basis and a different 'calculus' to the same effect, and hence presupposes what is in question in the present investigation. Goodman suggested that the perceptual units should be interpreted as basic, *neutral* elements that do not refer to any observer (Goodman, 1966, pp. 141–142). The neutral, "subjektlose" basis of the constructive system logically precedes any division into observers and objects alike. I do not doubt that for the technical purposes of construction the constructivist can abstract from the nature of the chosen basis up to a degree. Yet, we are entitled to an answer as to what the basic units are, and the only answer is: elements of a stream of subjective experiences. The considerations indicate that the hypothetical translation between any successful version of a sense-data based "Konstitutionstheorie" and the physicalistic counterpart theory would not be a routine mapping of terms. (This sets phenomenalistic constructivism apart from certain geometric and logical examples for systematically generating alternatives, which we will consider below.)

These points, I think, are well taken and provide reasonable doubt about the completablity of constructivism. However, the considerations fall short of *proof* that the constructive account of 'the external world', with the concomitant choices of bases, primitive predicates and calculi, is doomed to fail.[2] Even so, given this situation phenomenalistic constructivistic programs cannot serve as a serious argument for belief in strong underdetermination of any global theory of the world.

4.2 Things versus Numbers

In this section I am concerned with a particular kind of attempt to switch between one set of 'atoms' (individuals, basic objects) and another without disturbing the empirical consequences of a theory or global theory. This more modest argument preserves an essential feature of the phenomenological constructivism in generating systematically, and in an *a priori* fashion, conflicting alternatives. Competing choices of systems of atoms, or primitives, present a problem of sorts for realists, who tend to hold that our best theories settle in a unique way what is and what is not. The method in question is the so-called *pythagoreization* of scientific theories, i.e. the claim that theories about the natural world can in principle be reconstrued as to have natural numbers as atoms. The 'pythagorized' theories are *prima facie* empirically and perhaps rationally irresolvable alternatives to the orthodox originals. If one such construction succeeds, i.e. one mapping from physical objects to numbers exists, then many distinct mappings exists, and the way is paved to substantiate the inevitability of strong underdetermination U. There are three distinct ways discussed in the literature of how to systematically replace physical objects as individuals in our theories: by adoption of a *Koordinatensprache*, by proxy-function constructions, and by invoking theorems from model theory. Although all thinkers who considered an "ontology" of natural or real numbers for natural science reject it in the end, they reject it for widely different reasons.

---

[2]Compare on this matter the essays by G. H. Bird and Th. Mormann in Bonk (2001a).

1. *Coordinate languages.* Carnap developed the *Koordinatensprache* as an exercise in the construction of languages. To take one of his examples, for an object (a colour spot) placed in an abstract linear order of the progression type, '*Blue*(2)' may be be read as "The position having ... as coordinate is blue." where one would substitute for the dots in this case the number 2, or rather 0 of the progression.[3] More generally, assignments of measurable magnitudes to particles located at points in space-time can be translated in a similar way without loss of empirical content into assignments of 'corresponding magnitudes' to sets of ordered quadruples of real numbers. The quadruples are 'naively' taken as the coordinates of given, physical points of space-time (events), relative to a global, fixed co-ordinate system. However, one may drop reference to space-time points, or regions, as physical individuals, which are merely labelled or named by coordinates, and work directly with ordered sets of numbers as arguments of certain 'corresponding' predicates instead. (In general, of course, the frame of reference of the thing's motion has to be taken into account, but this is perhaps a mere technical complication.) This translation, relative to a given global co-ordinate system, is explicit, one–one and preserves empirical content.[4] The "ontological" reinterpretation of coordinate languages is due to Quine (Quine, 1981, pp. 16–17), compare Wilson (1981, p. 409). "Our physical objects have evaporated into mere sets of numerical coordinates" (Quine, 1976, p. 502).

Does the possibility of reformulating any theory in *Koordinaten-*

---

[3]See his *Introduction to Symbolic Logic* (Dover, 1958), section 40.

[4]Coordinate languages furnish an example of the *indeterminacy of translation* (Quine, 1977, p. 192). Carnap (ibid., section 39) distinguished several simple forms of the *thing language* according to the choice of individuals, with only suitable simultaneity and part-whole relations between the objects (these are not coordinate languages). A given thing at a point in time is considered as a "cross-section of the whole space-time region occupied by the thing". Conceivable individuals – primitive objects – for those languages are: four-dimensional space-time regions (i.e. things but not "thing-slices"); space-time regions of finite extent (i.e. things and their slices but excluding space-time points); all space-time regions; space regions of finite spatial extent (excluding space-time points); all space regions; or space-time points. Each alternative language form imports an alternative ontology from Quine's point of view.

sprache show the inevitability of empirical underdetermination? The coordinate language approach to constructing alternatives has a number of anomalies, one of them is the dependence on an explicitly chosen coordinate system. The dependence goes against the aim of present-day fundamental physics to discover (metric) invariants and absolute properties of the world. Coordinate systems are treated mainly as a computational help and technical convenience. Quine anticipated this objection. He noted that any coordinate-dependence shows up only in the statement of specific facts or particular state descriptions, but not in fundamental laws because they have coordinate free formulations. How can one explain this circumstance? Quine apparently accepted coordinate-invariance of fundamental laws as a happy accident. A more plausible and perhaps more favored explanation among scientists is that coordinate independence reflects a deep fact about nature, a fact that should be represented accordingly in our theories and in our philosophy of nature.

Another anomaly is that in switching to a coordinate language, together with Quine's ontological criterion, one ascribes physical predicates to abstract objects, like integers or ordered sets of reals. The suitably 'adjusted' predicates are peculiar (I wave the problematic inclusion of abstract objects into the scientific answer to 'what there is' and more direct objections to the applicability of the ontological criterion in this case (Wilson, 1981)). In the simplest case, like '*Blue*(2)', they contain explicit reference to the chosen coordinate system. The predicates are compounds of pure observation predicates and theoretical expressions. Thus, transformation to another coordinate system, say from polar coordinates to Cartesian ones, introduces another set of predicates, while *Blue* appears as blue throughout. There is good reason to single out predicates like color qualities, temperature, mass, momentum, etc. Coordinate systems are not a thread in the fabric of nature. Moreover, we expect causal relationships between at least some of an object's properties and its being what it is: the green color of a tree top, the electrical conductivity of a metal, etc. Abstract objects and their compound predicates hardly meet this requirement. And, after all, has all reference to bodies been eliminated in adopting a coordinate language? It seems not: for instance, Quine's proposal for temperature ascriptions in the coordinate language – "temperature in degrees centigrade of the object whose coordinates are" – has the word 'object' explicitly built into it. Defenders of the ontological interpretation have indeed felt uneasy about the consequent inflation of the predicates (Quine, 1976, p. 504; Putnam, 1975b, p. 184). The *Koordinatensprache*, in Quine's hand, is a "verbal regrouping" of the standard forms of the attribution of physical quantities to physical objects (homogeneous bodies, not systems) in terms of predicate logic. The coordinate language merely obscures the ontology of a theory, and physical predicates cannot arbitrarily be recasted. I will give this question more consideration a few paragraphs below.

2. *Proxies.* A "proxy-function", ϕ, is an explicitly given one–one mapping between two domains of an area of discourse together with a re-interpretation of the primitive predicates of the language, such that true sentences are mapped into true sentences. In other words, proxy-functions (the expression is due to Quine) generate *isomorphic models* of a theory. It is not required that the proxy-function is expressible with the means of either the new or the original theory (Quine, 1964, p. 218).

> The point is that if we transform the range of objects of our science in any one-to-one fashion, by reinterpreting our terms and predicates [...], the entire evidential support of our science will remain undisturbed. [...] The conclusion is that there can be no evidence for one ontology as over against another, so long anyway as we can express a one-to-one correlation between them.[5] (Quine, 1992b, p. 8)

Here is an example. Statements about particles located at certain points in space, say, can be translated outright into statements about

---

[5]Similarly: "Theories can differ utterly in their objects, over which their variables of quantification range, and still be empirically equivalent [...] We hardly seem warranted in calling them two theories; they are two ways of expressing one and the same theory. It is interesting then that a theory can thus vary its ontology" (Quine, 1992a, p. 96). See also the second part of Quine (1964).

natural numbers if a one–one correspondence between those particles and natural numbers can be established. Suppose there are denumerable many of (discernible, classical) particles and bodies build out of finite sets of these particles. Take one of the particles as the "first" one, call it $\mathcal{A}$. All other particles can then be put into a one–one correspondence ϕ with natural numbers by a simple, effective geometrical procedure that ensures that the collection of particles forms a proper progression (a first particle but no "last particle"; no repetition; every element can be reached from $\mathcal{A}$ in finitely many steps) at a fixed point of time. (I suppress the time dependence.) The basic physical predicates of the theory are adjusted accordingly to preserve the truth of all true sentences. For instance, we define for every monadic P a new corresponding predicate $\tilde{P}$ such that it denotes those and only those positive integers whose image under ϕ make the original predicate P true. Hence '$\mathcal{A}$ is positively charged' is true if and only if '$\tilde{P}(1)$', and so for all atomic sentences. Similarly for many-place predicates. For every, possibly quantified sentence in the original physical language there is corresponding one that is definitionally equivalent and quantifies of positive integers instead. The numerical alternative, "purged" of particles in favor of natural numbers, is *interdefinable* with its common rival but is observationally irresoluble (Putnam, 1975b, pp. 183–185).

Do truth-preserving *proxy* constructions show the inevitability of empirical underdetermination in our theories of the natural world? Putnam himself rejects proxy-constructions. He does so for a reason connected with the ontological ballooning of the 'proxied' theory's predicates, mentioned in the discussion of coordinate language: "a theory may presuppose objects as much through the predicates it employs as through the objects it quantifies over [...]" (Putnam, 1975b, p. 184). What the theory says about its entities through its predicates can determine the theory's true ontology. Since the remark quoted follows on Putnam's exposition of the example above, he might have had the following criterion in mind: if a physical predicate is true of some entities than these entities necessarily are material objects. This plausible criterion does not appear to be quite correct as it stands. The proxy theory assigns *arithmetical* predicates to integers, as part of its computational framework. By analogous reason then the integers represent mathematical objects, yet we know integers somehow also 'stand in' for particles in the original theory (Wilson, 1981, p. 420). This failure of the application of its arithmetical counterpart throws doubt on the initial criterion. It is easy to revise the criterion to escape the objection: if a theory assigns physical predicates to integers then the ontology of that theory comprises physical objects, besides whatever else. Physical predicates in a theory "flag" the commitment of the theory to physical objects in the theory's universe of discourse, however hidden they may be by the way the theory is formulated. One is then justified in giving preference to a formulation of the theory that makes this ontological commitment explicit. In any case, as plausible as the criterion in either version is, an unbiased 'judge' in matters ontological may take the proxy-theory with its apparently purely arithmetical ontology as a straightforward counter-example.

M. Wilson has advanced a more detailed diagnosis of what goes wrong when one accepts proxy-constructions at face value. Like Putnam he takes Quine's ontological criterion to task. Typical physical theories divide into a mathematical sector, the mathematical framework needed to state the theory, and a physical sector, reflecting what there is according to the theory. The proxy-theory, on the face of it, achieves a genuine "reduction" from a mixed ontology to one containing abstract objects only. In order to distinguish genuine from pseudo ontological reductions, Wilson proposed a set of requirements for when a reduction has *not* been achieved: (a) the alternative formulations are interdefinable (eventually by introduction of what Wilson calls "special constant axioms"); (b) the original theory's physical sector claims to represent the total ontology; and (c) its mathematical sector has a "normal" interpretation prior to and independently from the case under consideration. The three requirements are sufficient to rule out a proposed proxy-reduction as genuine. Any proxy-construction with characteristics (a) to (c) merely amalgamates the physical with the mathematical, creating a variant formulation that quantifies over for instance positive integers exclusively, without thereby ontologically (genuinely) reducing the one to the other. Thus existential quantification sometimes is not "univocal" (Quine) and does "double ontological duty", designating one sort of object in the physical and another sort of object in the mathematical sector.

Wilson's criterion carries immediate plausibility, but can it be defended on other independent grounds? The criterion yields the intuitively right verdict in Putnam's as well as in similar "proxy" constructions, for instance, alternative axiomatizations of classical point mechanics that apparently do away with forces (Wilson constructs analogous examples[6]). Yet, these are the very examples the requirements were abstracted from, I suspect. Since the correctness of those widely shared intuitions is precisely what is under question, appeal to intuitively right verdicts carries little weight. Taking proxy constructions' arithmetical ontology at face value has, of course, unwanted consequences, to say the least: it would make one believe, for instance, that forces in physical systems described by Newtonian mechanics do not exist. Wilson emphasises that, on the contrary, the notion of force is "vital" to physics. This is, I take it, a remark about heuristics in the practice of the physical sciences, with uncertain relevance for the ontological question at hand. After all, force as an element of a theory's ontology is hardly uncontroversial, as developments from Mach on show.

It is time to look for other arguments or approaches, and one does not have to look far. For Quine, see the quote above, proxy-constructions show that empirical evidence does not decide between ontologies thus related to each other. Note that he restricts lessons from proxy-constructions to applications to what he calls global theories, which makes examples like the pseudo-reduction of force-functions largely irrelevant. Extra-evidential factors, on the other hand, may weigh in on the decision and 'genetic' or historical considerations are part of it. Discourse about numbers *as entities* is not an autonomous form of discourse for us, as many writers have pointed out. It is dependent on prior mastery of the thing language, of objects or events located in space and time. The child cannot learn to refer to the abstract object 333 by ostension, in the way it can learn to refer to that

---

[6]J.C.C. McKinsey et al. (1953) and Wilson (1981, pp. 418–420) discuss axiomatizations of particle mechanics, where integers fully "replace" Newtonian force-functions assumed to be countable.

table over here or to Gaurisanker over there. Even if Quine is right that our "body-mindedness" or rather 'space-time-mindedness' is due to the contingent ways we have individually learned, or as a species evolved the language, it is still an irreplaceable stage in the acquisition of the lexicon of the speaker's language. (The procedure of establishing a coordinate system for scientific purposes presupposes the ability to refer to bodies.) So, working "from within" there is a good explanation for why we find it very hard to accept an arithmetical ontology.

One can do better, perhaps, by imposing an epistemologically motivated causality constraint on the kind of ontology permitted, or by emphasising the ontological continuity between directly observable bodies and theoretical entities. The first alternative excludes abstract objects in the image of proxy functions, since they have no causal powers and are not located in space-time and hence cannot exchange energy or momentum. This line may be criticized as *ad-hoc* move, as a repetition of the assertion that only things (or fields) in space-time are to be included in the ontology of a self-respecting scientific theory. The more promising second alternative faces the (Quinean) objection that ordinary bodies themselves are like all other objects "posits" by the human species, whose origins lie in dark prehistory. Hence the continuity requirement is satisfied under all proxy-functions and poses no effective ontological restriction. But are ordinary observable bodies (useful) theoretical posits? One may want to argue instead that one has a privileged epistemological access to bodies given in one's experience. Theoretical physical objects (genuine posits) in the sciences are essentially continuous with those immediately given objects. The argument for this claim has been stated in essence by E. McMullin in a classic paper on scientific realism. Consequently, replacing primitive physical objects by abstract objects becomes an untenable proposition, irrespective of the kind of predicates ascribed to them. Thus the second alternative has more of a fighting chance than the first alternative. It requires, however, the rejection of the naturalistic outlook, which made Quine's claim that observable bodies are posits plausible. For my purposes it suffices to have pointed out that the Quinean version of proxy-functions applied to global theories hinge on various additional epistemological and metaphysical claims (embodied in naturalism). Local applications of proxy-construction to individual theories, on the other hand, suffer from the uncertainties of confirmation theory and are open to the possibility that conjunctions with other theories undermine the proxies.

Two further remarks. Quine's premises include a hypothetico-deductive account of empirical evidence. Such a methodology is flawed (see chapter 3), but most commentators appear to share the view that in point of proxy-constructions of ontologies no alternative method of confirmation yields a different verdict. Since confirmation aims at estimating truth or falsity of (sets of) hypotheses, and truth values are invariant under the action of proxy-functions, hopes for a less gullible account of confirmation and scientific methodology to resolve the issue seem misplaced. There is much to say for this argument, but it rests on the assumption that it is not possible to separate the empirical support for a hypothesis and support for its existential presuppositions or implications except by way of the consequence condition (for the latter see Hempel 1966, p. 31). The assumption does not strike me as plausible and I am not convinced that it fits scientific practice throughout, but this is not the place to pursue this point.

Another premise concerns the idealization that the cardinality of material objects in our universe is countable infinite. This premise was one 'simplifying' element of Putnam's mapping of matter to numbers above. We have good reason to believe the premise is wrong and that the number of things, or primitive physical objects is finite (in the present context matter is viewed from a mechanical, pre-quantum point of view). The essential finiteness of matter in number has not been properly taken account of in ontological matters, and indeed it spells trouble for the application of the ontological criterion. Although this need not raise worries for the doctrine that reference is inscrutable, imposing a finiteness condition would tend to undermine the ontological impact of proxy constructions. The condition amounts to restricting proxies to mappings of matter into matter, i.e. (generalized) permutations, which may be taken to reveal nothing worse than *a priori* global semantic conventionality in our language.

(3) *Pythagoreization.* Pythagoreization is a label for the claim that the system of natural numbers is a potential "all-purpose" class of

individuals for (consistent) theories of physics and mathematics alike. A set of statements in first-order logic with a finite number of symbols, which has a model, also has a denumerable model, irrespective of what the intended subject matter is. As a consequence of the Löwenheim–Skolem theorem, there is a denumerable model for that set of sentences (the theory) whose domain are the positive integers. For mathematical theories, like axiomatizations of the real numbers or ZF set-theory, this fact has created some consternation since these theories were intended to capture truths about non-denumerable sets. Our epistemically best *physical* theory of the material world taken together with all true observation sentences is a presumably consistent set of statements (with a finite vocabulary), has a model, and so the theorem applies. If the epistemically ideal theory satisfies additional formal conditions, then under the "numerical" model the primitive predicates of the theory have an interpretation as proper arithmetical predicates, definable in terms of logic, product, sum, etc. This though is the uninteresting case, since the reinterpreted theory has lost its empirical content. The stimulating claim is that there is a numerical interpretation of the ideal, total theory which preserves its empirical content (and even takes our perceptions into account).[7] Quine summarizes:

> This theorem does not, like proxy-functions, carry each of the old objects into a definite new one, a particular number. [...] Despite this limitation, however, the reinterpretations leave all observation sentences associated with the same old stimulations and all logical links undisturbed. Once we have appropriately regimented our system of the world or part of it, we can reinterpret it as to get by with only the slender ontology of the whole numbers; such is the strengthened Löwenheim–Skolem theorem [a version due to Hilbert and Bernays – T.B.]. (Quine, 1992a, p. 33)

---

[7]A *locus classicus* is Putnam's Presidential Address before the Association of Symbolic Logic "Models and Reality" 1977, reprinted in Putnam (1983b, pp. 1–25); compare Quine (1976) and "Ontological Reduction and the World of Numbers" (1964) (Quine, 1966, pp. 212–220).

The theorem thus appears to provide strong support for the necessity and universality of the observational underdetermination of theories. Although the theorem states that a denumerable model exists, provided the theories in question meet certain formal conditions, it does in general not provide a specific, explicit correlation, one–one or one–many, between any physical object in the domain of the initial theory and a number (Quine, 1964, p. 219). A proper reduction of (primitive) physical objects to integers, say, require *explicit correlations*, otherwise it has no explanatory power and is all but *pro forma*. The standard instances of reductive relations between theories, like the identification of temperature and mean kinetic energy in the statistical mechanics of gases for instance, have this characteristic. This constructive shortcoming separates the theorem from the more "pedestrian" proxy-constructions considered on the previous pages, as well as from examples like Goodman's plane, Whitehead's construction of points as nested sequences of "balls", etc. An empirically equivalent rival theory, whose existence is logically guaranteed but whose formal features and reductive relationships are not fully explicit, do not undermine the justification a theory derives from its verified empirical (hypothetico-deductive) content. It appears that on this ground Löwenheim–Skolem's theorem cannot provide a strong basis for the claim that empirical (and rational) irresolubility is inevitable.[8]

I like to add another thought. A paragraph back I have suggested questioning the assumption that one can idealize the material content of our universe as (denumerable) infinite. The assumption is explicit

---

[8]Putnam has given the 'constructivity requirement' another twist in his discussion of the Pythagoreization issue (Putnam, 1977, pp. 23–25). He proposed to adopt a non-realist semantic theory: actual community-wide usage, in the metalanguage, 'fixes' semantic relationships and determines that "cat" refers to cats and not to dogs, etc. Once 'unintended' interpretations of a given object-language (our global theory pythagorized) are made explicit they can be ruled out as false, as conflicting with current usage. Hence Putnam's remark about models not being "lost noumenal waifs looking for someone to name them" (ibid., p. 25). This approach is reminiscent of Horwich's *ansatz* discussed in section 5.3.

in Putnam's proxy construction, based as it is on a one–one correspondence of (classical) particles and positive integers. Similarly, if the physical individuals of the "global theory" of the world (in Quine's sense) are finite in number, then Löwenheim–Skolem's theorem packs little ontological punch. Models of the physical sector, as I called it, of said system must admit finite domains. However, the domains of the theories that together make up the standard mathematical framework, ultimately ZF, have no finite models. I take the finiteness of matter, understood as the bounded number of physical individuals in the universe, as a fact and as an essential characteristic of material systems. Does this fact allow us to block on principled grounds arguments for the inevitability of empirical irreducibility of the kind we have considered in this section?

The notion of finiteness of matter needs some clarification. From a physical point of view: has the notion been subverted by the growing importance of a field-theoretic point of view in the fundamental sciences? The relation between particles and fields is notoriously difficult to pin down; surely discreteness versus continuity is one issue, finiteness versus infinity is another. A field, extended through space at any one instance, has a finite energy content and is better viewed as one individual than as a non-denumerable collection of physical states, each defined for every space-time point. And neither photons nor virtual particles count towards a proper material ontology, nor does the fact that the number of particles is not an invariant in relativistic collisions of elementary particles devaluate the validity of the concept of finiteness. The problem of the infinite divisibility of matter that was so much on Kant's mind in the discussion of the antinomies or pure reason and in the *Anfangsgründe der Naturwissenschaft*, appears to be settled by modern physics in favor of the finite doctrine. I conclude from this brief review that there is nothing in the exact sciences that definitely would contradict or undermine the usefulness of the notion of the numerical finiteness of matter. Phenomenological finiteness is pervasive in our environment: one sees and interacts with finite collections of things, and counting and computing are similarly restricted to finite operations. Yet, an entirely unproblematic statement of the general concept of finiteness is hard to come by. The intuitive concept "finite set" cannot be fully captured (axiomatized) in terms of first-order predicate logic. Dedekind's definition derived negatively from his well-known definition of an *infinite* set. Second-order logic, however, offers a natural and unique definition of finiteness: a set is "finite" if every one–one function on the set has every element of the set in its range. But second-order logic faces certain objections: it is ontologically "inflationary", and its adoption as a framework looks like a desperate *ad-hoc* move. Th. Skolem and recently S. Levine (1994) have stressed that a notion of finiteness is conceptually prior (if not strictly *a priori*) to any proposed formulation of finiteness in set-theoretic terms. The very notion of mathematical "proof" requires the idea of a finite number of steps or premises (infinitary systems of logic notwithstanding).

Quine, however, claimed that to ascribe an ontology to finite systems of matter is *meaningless* and inconsequential, when done from the vantage point of the theory in question. This is a somewhat startling claim (Quine, 1969b, pp. 62–64; 1964, p. 216). I shall argue now that this claim says more about Quine's use of the terminus "ontology" than about ontology. There are two reasons for Quine's assessment, the more general of which has to do with the doctrine that the specification of an ontology is a relative affair, i.e. it is meaningful only relative to a translation manual or a sufficiently strong background theory. I do not want to dispute this controversial doctrine at present (but compare section 6.2), and rather address the question why the ontological assessment of finite systems, when viewed theory-internally, would "loose all force". The reason given is that with the replacement of quantifiers by chains of conjunction and disjunctions the variables disappear and so there is no sense in asking what they range over (Quine, 1969b, pp. 62–64). (Provided the finitely many individuals are named, which is a difficult practical requirement to meet if the numbers are large, the individuals are not observable or indiscernible.) Suppose the speaker lives in a small-scale world composed of just a few things, which make true a couple of observable predicates like hot, green, heavy, slow, etc. Under these circumstances the speaker's most simple and economical systematization of her experiences would perhaps be one that does not refer to things at all, but only to recurrent patterns of experiences. For instance, the speaker may have acquired separate dispositions to associate each of distinguishable raven patterns, no. 1 to no. 13 say, with the experience of a black, slim silhouette. No grammatical referential apparatus is required to be able to identify across space and time, generalize and predict in such a small-scale finite world, at least if one follows Quine's (speculative) picture of the process of reification. However, why should one want to devolve the quantifiers? Given that scientists are working with sets of individuals up to the order of 10^{80}, and power sets thereof, it seems definitely preferable to keep the quantifier notation in place although the range of the variables is bounded. Most of these individuals will remain unobserved though perhaps observable. I anticipate the objection that in a small-scale world posits of "unobservables" are (logically) superfluous for purposes of prediction and control, if not from a point of view of ontological economy. The objection has a certain force from a traditional empiricist standpoint, but not from the naturalist's. The naturalist needs fully to take into account the human cognizer's bounded resources (as Quine took them into account in his examination of Craig's theorem, compare section 2.2). That means the quantifier notation is an essential instrument in managing large numbers of physical objects although the numbers in question are strictly finite. Finiteness in number is part of what defines material systems and separates them formally from systems of abstract objects that serve as domains for number theory and analysis (for instance, Peano arithmetic has no finite models).

To turn back to the question posed at the beginning of this section: what does Pythagoreization show about the inevitability of empirical underdetermination? Very little. The lack of explicit, constructable correlations between the system of primitive material objects and the set of integers, say, and taking full measure of the difference in cardinality between material systems and progressions and the like tend to diminish the importance of the possibility of a 'pythagorized' material world as an argument for global underdetermination.

4.3 Squares, Balls, Lines, and Points

In this section I consider a family of suggestive examples drawn from geometry that were intended to illustrate mechanisms by which empirical underdetermination could be established. They aim to lend credence to what I have called the "universality thesis" of underdetermination: there is not one significant physical property of the basic objects a theory postulates, not the objects itself, that would remain invariant in all empirically equivalent alternatives, and underdetermination is inevitable. The relativity, or 'amorphousness', or denotational vagueness that besets geometric notions like points, lines, regions in their respective theories carries over, it is claimed, to physical objects and events. "Physical objects and events and perceptual phenomena go the way of points and lines and regions and space" (Goodman, 1978, p. 119). Putnam, for instance, in arguing against metaphysical realism, uses the Whiteheadian elimination of points (see below) to illustrate that observationally indistinguishable scientific theories can substantially differ in their respective ontologies (Putnam, 1983a, p. 42).

> All this isn't an artifact of my simple example: actual physical theory is rife with similar examples. One can construe space-time points as objects, for example, or as properties. One can construe fields as objects, or do everything with particles acting at a distance (using retarded potentials). (Putnam, 1978, p. 133)

How serious should a would-be realist take this kind of argument for the underdetermination of any substantial theory? A. N. Whitehead (1955) advocated the "elimination" of points from the physical geometry of space. Actual testing and measurement in the sciences probes finite regions of four-dimensional space-time. Hence, bounded spatial regions are empirically accessible physical objects while points are not so accessible. From an empiricist point of view statements referring to points, like an ascription of a field strength, are of questionable cognitive significance. In order to put science on a more rigorous basis,

Whitehead identified, in best constructivist tradition, points as limits of a nested sequence of unit balls (but other kinds of open, bounded spatial regions do as well). Defined in this way, sentences referring to points can be translated into complex sentences about finite regions, unit balls, or converging sequences thereof. The notion of a point is still retained as an "ideal" element for technical purposes, or as a notational shortcut, but not as a genuine theoretical object on par with the earth's axis of rotation, say. Whitehead rejects the notion of an instant of time as well, but for the purpose at hand I consider space separately.

In considering Whitehead's proposal for the "elimination" of points from space, two questions need to be confronted: Does the construction succeed in generating an alternative ontology? And if it does succeed, is the alternative ontology fully compatible with modern physics, the area of science most directly involved, in evidential or heuristic respect? (I will not question Whitehead's philosophical motives in introducing the construction.) The answers depend, first of all, on granting space ontological status as an entity in its own right, besides particles, forces, fields, etc., that is on adopting a broadly substantialist view of space or space-time. The answer further depends on whether it makes sense to say that physical space has natural 'elementary parts' at all, be they points or balls. If choosing the one over the other in representing physical theories resembles more a cutting-up of the same pie in different ways, if space is ontologically amorphous, the construction is not an instrument of a proper ontological reduction and cannot underwrite underdetermination.

Although the purely geometric notion of an affine space, say, explicitly involves the notion of a set of points (together with a set of vectors, which has the structure of an n-dimensional vector space, and a unique displacement vector for any two given points is required to exist), one may grant that the formal construction achieves its aim in a Euclidean space and locally in non-Euclidean spaces. It remains to be shown whether it does so *globally* in non-Euclidean spaces. A reason for doubt is that in this case the construction has to take the metric or curvature (connection) into account, which is usually defined for every point of the manifold. Note, the limit-construction needs to be extended ('lifted') to all functions defined on the space, i.e. the functions, like electro-magnetic fields, are then primarily defined on open bounded regions, say, and the limes to any given 'point' is required to exist.

Second, even if the Whitehead construction succeeds, it does not appear to lead to a fully adequate reconstruction of physical theory. There is evidence (for instance, from the behavior of electrons and quarks in high-energy scattering experiments) that some objects have a genuine point structure. Besides contradicting the naive premise that experiments can only probe finite regions of space, the experiments provide good inductive grounds for accepting point-particles and hence for points as physical objects. A similar lesson can be drawn from the explanatory necessity of positing point-singularities of fields and space. It is a tell-tale sign that Whitehead retains points of space as "ideal" elements for formal purposes, i.e. for easing the statement of the theory and the manipulation and application of its theorems. Yet these very reasons indicate that those ideal elements are in practice indispensable, and are separated from proper theoretical entities only verbally, by way of attaching the label "ideal", and not by way of function in the overall edifice of the theory. In sum, despite being frequently quoted, Whitehead's construction gives only weak support, if at all, to the thesis of a general underdetermination of our theories, particularly in its strongest version.

Geometry has always been considered a rich source of examples for how truths can *underdetermine ontology*. Hilbert's familiar axiomatization of Euclidean geometry has *six* primitives: point, line, plane, and the relations incidence, betweenness and congruence. The independent primitives 'line' and 'point' of Hilbert's system of plane geometry can be replaced in several ways and without recourse to infinite sequences. One may choose, for instance, to eliminate points in favor of the second primitive, lines by defining a "point" of the plane as a special pair of lines Goodman (1978). In the geometrical system S_P, where a complete set of parallels P along with its orthocomplement is given, points correspond to a unique pair of lines, one drawn from P, the other from its complement. Geometrical theorems which involve reference to points in the original (Hilbertian) geometry translate into theorems referring to lines in S_P; in other words there is a structure preserving mapping between 'pair of lines' and 'point'. The construction varies with the choice of P, but the systems S_P, $S_{P'}$, ..., do not differ among each other on what there is. Conversely, point sets can consistently replace the primitive "line", generating a geometry S_T. (I refer to this example as *Goodman's plane*.) The Whiteheadian method of construction (S_W), relying on a more complex conceptual scaffolding, finally yields yet another alternative ontology of the Euclidean plane. S_P and S_T (and S_W) differ in their respective "ontology". All domains are on par with regard to the original set of geometrical truths. When Goodman claimed that these representations of the plane represent "antagonistic worlds" he does not want to re-emphasize the well-known fact that a consistent elementary theory has more than one model. Rather, the claim is that the geometries S_P and S_T of the plane are in genuine conflict, i.e. they cannot possibly be reconciled (in one "world") (Goodman, 1978, p. 9). These geometrical systems are best understood as separate yet equally "real" worlds in their own right. In order to demonstrate the conflict between representative systems Goodman exhibits two statements s_P and s_T (actually statements in a common meta-language). Each one is true in one but not in the other geometry. Their conjunction is true in none of the geometries.

s_P Every point is made up of a line from P and a line from its orthocomplement.

s_T No point is made up of lines or anything else.

Goodman's critical discussion of how one might want to resolve the conflict is instructive. The time-honoured strategy is the explicit conditionalization of each statement to one geometrical system, and to replace the stronger notion of compositionality (line "composed" of points, etc.) by the notion of correlation. One arrives at statements like: 'Under system 1, there is a particular combination of lines, one drawn from P and one from its complement, *correlated* with every point.' The relativized pair of statements, however, does not say the same thing as the original pair s_P and s_T; they are ontologically neutral but also significantly weaker. The realist's question 'What are the primitive objects of the plane?' is not answered by making a list of the kind: 'In framework 2, they appear as points.', 'In framework 3, they appear as an infinite sequence.' The new readings of s_T and s_P, relativized to a specific system and with correlation replacing composition, "say nothing about what makes up a point" (Goodman, 1978, p. 116). Reconciliation becomes possible only by surrendering all significant differences between the geometrical systems.

> A world with points as elements cannot be a Whiteheadian world having points as certain classes of nesting volumes or having points as certain pairs of intersecting lines or as certain triples of intersecting planes. That the points of our everyday world can be equally well defined in any of these ways does not mean that a point can be identified in any one world with a nest of volumes and pair of lines and triple of planes; for all these are different from each other. (Goodman, 1978, p. 9)

In whatever way one interprets these reflections on the ontology of geometry they fail to give convincing support for the underdetermination thesis. *First*, one may read Goodman's reflections in the light of applied geometry, i.e. in view of the role of the geometry of space for the kinematics and dynamics of physical objects. Assuming that purely 'spatial entities' count towards the overall ontology of a physical theory, Goodman's plane (or space) is, on this account, intended to show that neither observation sentences, nor natural laws nor geometrical theorems fully determine the theory's ontology. S_T and S_P are empirically indistinguishable, therefore the theory's overall ontology is empirically underdetermined. On this account, Goodman's considerations are of one piece with the earlier ones by Whitehead – and they fail for the same reasons.

Second, one may read the previous reflections on the ontology of geometry as setting out an example, the essence of which can be generalized and transferred to the material ontology of physical theories, or one that establishes a distinct conceptual possibility, which applies to physical objects as well as to geometric ones. Judging from the quotes by Goodman and Putnam in this section, this seems to be the interpretation intended (compare Putnam, 1978, p. 132). Yet the relationships between points and lines on Goodman's plane have features that make straightforward transference to typical relationships between the sort of entities postulated by physical theories all but impossible. One may indeed define points by sets of lines or vice versa, but hardly electro-magnetic fields by sets of point charges or vice versa, leptons by hadrons or vice versa, etc. The Wheeler–Feynman theory of electrodynamics does not prove the contrary, see chapter 2; but even if it did we would have found an isolated example: not enough to show the inevitability and all-pervasiveness of underdetermination. Furthermore, S_T and S_P are strictly interdefinable (Wilson, 1981), but interesting actual and historical examples of empirically equivalent theories, like 'versions' of quantum mechanics or the Wheeler–Feynman theory, are not so interdefinable. The variant formulations of Goodman's plane have all the characteristics of what P. Horwich called 'potential notational' variants.

There is a further consideration that thwarts drawing far-reaching lessons from the geometric case *per se*, namely the need to prove the consistency of the defining axioms of Goodman's plane and other, richer systems of geometry. The (relative) consistency of these geometries is demonstrated by way of a model in the real numbers. This is done by a coordinatization of the plane, i.e. by interpreting a point as an ordered tuple of real numbers, and a line as a linear equation. The need to have recourse, for the demonstration of consistency, to a model that has neither points nor lines, say, as primitives, indicates that worries about ontology are misplaced in this case. Indeed, Goodman's plane represents a kind of case where rational irresolubility – with respect to the geometric theorems as data set – of conflicting "ontological claims" is plausibly interpreted as result of an underlying *indeterminacy* (compare section 6.4). The correct view, I believe, is that there is no autonomous ontology to be committed to in geometric systems like these. Considerations as such have persuaded some thinkers to abandon mathematical realism outright and others to adopt a structuralist view of mathematics.[9] Any such view sets mathematical theories apart from physical theories about the natural world, as they are typically interpreted by scientists and scientific realists (though not all), and thus undermines the alleged generalizing and illustrative power of the geometric case.

4.4 Algorithms

The existence of *algorithms* for generating observationally indistinguishable theories would surely prove the claim of inevitability of empirical irresolvability. In this section I examine four proposals for generating alternatives to any given theory. The proposals are somewhat less controversial and more conventional then the semantic reinterpretations considered in the previous section. Methodological criteria of admissibility and rationality of theory choice aside, below I argue that none of the algorithms proposed qualify as supporting thesis U (p. 19) of empirical underdetermination.

Like many ideas in connection with inferentially indistinguishable theories the first stratagem to be considered has its origin in empiristic views on cognitive meaning and testability. The prescription for generating alternatives is rather simple: form the conjunction of the empirical theory in question with any "non-factual", descriptive sentence that is not analytic. Following Otto Neurath and Carl G. Hempel we call an *isolated sentence* a sentence that is neither a logical truth nor a logical falsehood, nor a sentence that has any experiential relations. An isolated sentence (by definition) can be conjoined with a body of theory without increasing the theory's scope or explanatory power. Their conjunction has exactly the same deductive empirical content as the original theory, and by making suitable choices for those claims

---

[9]S. F. Barker, for one, has plausibly argued from the existence of conflicting set theories against mathematical realism, in J. J. Buloff (ed.) *Foundations of Mathematics* (Berlin 1969), pp. 1–9. Versions of structuralism in mathematics have been advanced and defended by B. Hellman, M. Resnick among others. For Quine's take on structuralism in the natural sciences compare Quine (1975, p. 9; 1992a, p. 96) and Quine (1992b).

one can construct arbitrarily many conflicting alternatives. A common misconception is that isolated sentences are necessarily "metaphysical" sentences. Hempel provided a counter-example with a bi-lateral reduction sentence, devised by Carnap as a way of partially interpreting theoretical predicates. Moreover, the property of being isolated is theory or *language relative*: a sentence is isolated in one theory, yet is non-isolated and perfectly admissible in a richer theory.

The "conjunction"-algorithm for generating alternatives is commonly dismissed by appeal to a principle of conceptual economy (Ockham's razor, Friedman's "parsimony") or by pointing out that the empirical support of the conjunction does not extend to "idle" hypotheses (Wright, 1992, p. 25). An isolated sentence can be eliminated without loss of scope and explanatory power. Both reasons, although basically correct I believe, need certain provisions. One can construct two *logically equivalent* theoretical systems, one of which does have an isolated sentence among the set of its primitives sentences, and the other does not. "Primitive" sentences are those sentences in a reconstruction of a theory which are built from a primitive vocabulary, using the logical apparatus, and which are not derived in the theory from any other sentences. Suppose S_1, S_2, S_3 is an enumeration of a theories' non-logical primitive sentences, and S_3 is the isolated sentence. Consider a theory in which the same set of primitives is given by $S_1, S_2 \wedge S_3$. Both theories are logically equivalent, but the latter truth-functional compound is not an isolated sentence (Hempel, 1951, pp. 115–116). Difficulties like this one can be overcome (for instance, by restricting what qualifies as a primitive predicate), but they need to be addressed with care by critics of the procedure. Nor is wielding Ockham's razor a routine matter. As Hempel explained in "The Theoretician's Dilemma": the rule can be applied so as to eliminate, along with any isolated hypothesis, the whole theoretical apparatus of a theory, leaving only the theory's Ramsey sentence to express its empirical content. In any case, the "conjunction"-algorithm clearly does not prove any strong underdetermination claim and cannot underwrite anti-realism.

Next, I turn to a recent proposal for generating empirically indistinguishable alternatives, which introduces an *observer-dependence*. Let the given theory be τ_1 and let τ be a theory that is incompatible with τ_1 on the same domain. Define an alternative τ_2 to τ_1 by the following stipulation: (i) the laws of τ_2 are those of τ *as long as no tests* (or observations) on τ_2 are made; (ii) the laws of τ_2 are those of τ_1 whenever the theory is actually tested. The original theory τ_1 and its bizarre alternative τ_2 are empirically irresoluble and incompatible; the construction is quite general (Kukla, 1993, pp. 4–6). The proposal supports the strong universality thesis, i.e. for every theory there are empirically equivalent ones that conflict with it on every statement chosen. (The algorithm is reminiscent of Reichenbach's construction of an "egocentric language" with its characteristic breakdown of ordinary causality and observer dependence.)

The degree to which the "observer"-algorithm (and similar theoretical concoctions) strains common and scientific sense needs no elaboration, but what exactly is wrong with it? For one, it ignores the possibility of *continuous observation* of all relevant observables. Admittedly, only sufficiently simple systems permit systematic continuous monitoring, except under highly idealized assumptions. In the methodological vein, Laudan and Leplin in particular have suggested criteria to separate the wheat from the chaff in matters theoretical, like 'being non-parasitic' or 'being taken seriously by the scientific community'. Each has initial plausibility, yet for the tricky purpose of excluding theoretical rivals *a priori* none has been shown to be properly grounded in considerations of truth-conduciveness or testability.[10] For instance, a time-honoured methodological criterion to dismiss a theoretical rival is its lack of simplicity in case no evidence tips the balance. It is true, in a charitable way of speaking, that the algorithmic rival is less simple than the theory on which it builds. The difficulty in wielding the simplicity criterion is that we do not have a firm grip on the very notion of absolute or comparative simplicity to start with, and, more

---

[10] I agree with Kukla's assessment that the criteria proposed by Laudan, Leplin and others are either question begging or do not effect a real selection at all; see the exchange of papers between the authors and Kukla on this issue, summarized in Kukla (1998, pp. 66f). The proposals considered below, however, are not discussed therein, neither is the proposal I make in connection with tempered pluralism in section 1.2.

importantly, simplicity's relation to truth and falsehood is tenuous in general, at least outside the domain of statistics and probability theory. The simpler of two alternatives seems no more likely to be true, if one goes by past experience with simple hypotheses and scientific practice – even if one correlates 'simpler' with being 'more unified' or more 'coherent' in a Bayesian way. Both difficulties have to do with the fact that simplicity rankings of theories are sensitive to the scientific languages chosen for expressing a theory's hypotheses, and that the notion of simplicity cannot be fully characterized in the abstract.[11]

There are other epistemic virtues to reckon with in a theory. τ_2 generates regular predictions through τ, which are not testable in principle due to the manner in which the theory is constructed. The only way to refute τ_2 is by refutation of its sub-theory τ_1. In point of direct refutation then there is a suspicious looking methodological asymmetry between sub-theories τ and τ_1. For instance, one would hold it against classical electro-dynamics if the theory were only testable in the domain of electro-static phenomena through its corresponding sub-theory, and not independently in the domain of induction phenomena as well. It is not as if τ_2 is less refutable than τ_1, rather τ_2 *should be more testable* on account of its content associated with τ than it actually is on account of its particular construction. (The consideration does not, of course, show that τ_2 is false.)

Another line of objection is that τ_2 generates *virtual predictions* which, although untestable, are false most of the time. Therefore rejection of the contrived alternative appears fully justified. This line depends on broadening the "test basis" and the empirical content to include contrary-to-fact statements. The broader conception of empirical content, however, goes against the spirit of the very notions empirical content and test basis: collections of facts or observable states of affairs make uneasy bedfellows with conditionals like "Had you measured the field-strength at this spot 30 minutes ago, you would have found it to be zero". Its truth-value depends on the presumed truth of the relevant theory (and the idea of natural laws), though this is

---

[11]Kukla argues in detail that this algorithm and a variant cannot be selected on the grounds of a lower ranking in a simplicity ordering (Kukla, 1998, pp. 77f).

easily forgotten, when the theory in question is very mature, or common sense contrary-to-fact conditionals are at issue. Yet, in some cases its truth-values will be determined by background information and independently from the theory or its rivals under consideration. This may remove underdetermination in selected cases, which qualify, but it is not a universally available prescription. Moreover, some thinkers, Quine and Goodman among them, find the semantics of contrary-to-fact conditionals from a more principled philosophical point of view highly suspect; lack of consideration of this point hampers Laudan and Leplin's criticism (LePore, 1986, pp. 12–13).

Promising among other responses (see the discussion of the evidential role of background-information in section 3.5) seems to me also the following reflection: given the theoretical rivals generated by the observer-algorithm it is a 'strange' fact that there are empirically successful theories, which do without this kind of extra. In whatever area of research one looks, there are perfect examples of empirically successful theories, which tell a coherent causal story and have no observer-dependence of the kind suggested.[12] This fact requires *explanation* and the requirement does not apply symmetrical to underdetermined algorithmic rivals. The simplest explanation of the success of straight theories is that nature is structured in a way that makes introduction of an observer-dependence superfluous: the laws of τ_1 are approximately laws of nature, very roughly put. The explanation of the empirical success of τ_2, on the other hand, is too easy to come by: the success is guaranteed *a priori*. So, one has an epistemic reason to believe in straight theories, although it is logically possible that they are false and the algorithmic rival is true. (Note, modern micro-physics, in common understanding, is a theory that displays observer-dependence and has a-causal features. The crucial, structural difference is that the observer-dependence of the values of observables in quantum mechanics is law-like. The observer-dependence in our *Ersatz*-theory is unpredictable and at will, at bottom subjective.)

Although I have discussed only one of many possible ways of

---

[12]Laudan and Leplin argue on the basis of this observation inductively against the algorithmic rival in Leplin and Laudan (1993, p. 13), but see section 3.5.

constructing rivals to a given theory, the general conclusion must be that a reason for believing in strong underdetermination should rest on a more solid foundation.

A variation improves on the previously considered algorithm by getting rid of the subjectivistic dependence on the act of observation. The strategy retains a positional (indexical) dependence and the characteristic a-causal switch from one set of laws to another. The "hypothesis" is that the world would have been governed by a set of entirely different natural laws, or none at all, from the moment a particular person, Socrates say, had uttered a certain word at a certain moment in the past (alas, he didn't utter the word).[13] Any system of belief that entails this hypothesis is manifestly inferentially irresoluble from the common sense system that entails the belief 'if Socrates had uttered the word at that time it would not have had any effect on the level of natural law'. Any number of such systems can be constructed in the same fashion. The strategy, however, does not establish the inevitability of observationally underdetermined beliefs sufficiently strongly to undermine realism in the sciences. For it postulates or grants one set of laws and unobservable entities before the contra-factual event and another set afterwards. Hence for the question whether *there are* unobservable entities in the first place, and whether statements about them are true or false, the algorithm is irrelevant. The alternative is compatible with the semantic component of scientific realism, and perhaps even with its epistemic component. Moreover, any hypothetical taker of such a system of belief is exposed to the charge of incoherence. For, by accepting the system of scientific theories before the crucial event and after, he already has ample evidence that natural laws do not depend on individual acts. So inductively he should conclude that the bizarre contrary-to-fact conditional above is actually false.

A somewhat specialized parameter-algorithm for generating inferentially irresoluble alternatives utilizes the dependence of scientific theories on a number of physical *parameters*.[14] For instance, the viscosity

---

[13]The idea has been attributed to Putnam. I suspect the intend was to underscore the dangers of neglecting contrary-to-fact conditionals, as Quine did, for discussions of empirical equivalence, compare Putnam (1975b, p. 180).

[14]Newton-Smith (1978) and Wilson (1981) address the question of ϵ-

of a fluid, the rest mass of a fundamental particle, the velocity of light in vacuum, etc. The values of these parameters are not usually explained in any sense by the theory in question, although sometimes an application of the theory can in an interesting circular way help determine those values. A theory's observational consequences, in general, vary smoothly with the value of these parameters. Consider now the substitution of the numerical values of a given set of parameters (or any subset of it) by values that are numerically "close". The result is a theory, based on a new set of parameters, whose observable consequences are indiscriminable, by current experimental methods, from the observable consequences derived based on the original set of parameter values. Provided, of course, the choice of the new set of values is numerically "sufficiently close" to the original one. The existence of a plethora of secondary theories in this sense is a mathematical truth whenever there the observable consequences depend smoothly on the parameters.[15] (Given the known non-zero margin of experimental error of our currently best methods and instruments, one can calibrate new sets of values such that any in principle observable differences in prediction will be swamped by the measurement error.) This ϵ-construction guarantees both incompatibility and empirical equivalence of primary and alternative theories. Another ϵ-construction, this time not applied to a theory's parameters but to space and time, yields a similar result. Hermann Weyl already had devised, with the aim to bypass commitment to the continuum, a plane geometry that recovered the usual theorems on the basis of rational numbers. The ϵ-construction goes beyond that in that it replaces the smooth space-time manifold by a grid structure in the context of classical mechanics. The feasibility

---

constructions or parameter-algorithms. The possibility was known for much longer in a different context: minute differences between numerical assignments to quantitative terms like "length in centimeters" pose a difficulty for an operational account of the meaning of theoretical terms. If operationally indiscriminable, should two such statements be assigned the *same cognitive meaning*? (Hempel, 1966, pp. 110, 129).

[15] The restriction to theories with smooth dependencies is necessary. Many physical systems exhibit significant qualitative changes if one of the relevant parameter takes on, say, a non-integral instead of the original integral value.

of a finite difference version of Newtonian particle dynamics with an undetermined grid-distance parameter h restricted to the rational numbers as an observationally underdetermined alternative is examined in Newton-Smith (1978, pp. 102–104). The 'primary' theory is the one with $h \to 0$. The 'secondary' theories, based on small values of the grid-parameter h, are shown to be as empirically adequate as the primary theory.

Do ϵ-constructions then make the inevitability of empirical underdetermination, and the universal scope of conflicts of assertion, plausible? The first kind of example, in opposition to the second one, is surely trivial, but that alone cannot justify its rejection as an instance of empirical irresolubility as Wilson (1981, pp. 209–211) has agued.[16] What limits its effectiveness are two features: (a) natural laws are not usually individuated by the values of parameters. We tend to think that Newton's law of gravitation, say, is the same law whatever the value of the gravitational constant (as long as it is positive and nonzero). Hence alternatives based on ϵ-constructions do not conflict on the laws of the primary theory and not on observational consequences either. Whatever conflict of assertion remains is strictly localized to incompatible assertions about the values of certain invariant quantities. This class of examples then does not underwrite the universality claim of irresoluble conflict. (b) The empirical equivalence between primary theory and secondary parameter-alternative is a matter of ignorance. Let us distinguish between (i) a theory "kernel"; (ii) the auxiliary hypotheses necessary to model a physical system and to arrive at predictions and explanations; and (iii) the auxiliary hypotheses necessary to describe the actual measurement.[17] The claim to empirical equivalence between primary and ϵ-alternative relies on the conjunction of hypotheses from all three levels, apart from extra initial data.

---

[16] Primary and secondary theories turn out to be empirically equivalent even on Wilson's "inter-active" notion of observability and observational equivalence, inspired by the measurement problem in quantum mechanics (Wilson, 1981, pp. 220–221, 226).

[17] My coinage of a theory "kernel" differs from Lakatos' well-known notion "metaphysical core" in not including the metaphysical parts, and corresponds roughly to Wilson's theory "core" (ibid., p. 212).

The theories' direct observational consequences, however, stem from conjunctions of levels (i) and (ii) only. Primary and secondary alternatives are *not* equivalent relative to this set of quantitative observational consequences. The claim to observational equivalence then refers explicitly to what we can or cannot measure. This presents a significant departure from cases considered earlier where empirical contents of rival theories, defined on the levels (i) and (ii), were proven and known to be identical. Moreover, since *in the limit* of exact, error-free measurement all but one of the rival alternatives are refuted (provided one is correct at all), I think one is justified in claiming that there is a fact to the matter. ϵ-constructions fail to support the theses which are of interest here.

Turning finally to the *discretization* of space and classical particle mechanics, the second example in this family, it too has features that make it an unfit instance of observationally irresoluble conflict of assertion. (a) The discretized secondary theory fails to be an autonomous alternative to the primary theory ($h = 0$). The finite difference version of Newton's force law does conflict with the original statement. This conflict however is not one of outright contradiction but a numerical matter of degrees of approximations. The method of discretization, quite apart from reasonable questions about the applicability to proper space-time or quantum theories, thus fails to generate alternatives which challenge each and every law of the primary theory. (b) As long as the distance parameter h is non-zero, observations in the limit of error-margin free measurement can discriminate between secondary and primary theories. And, as argued above and earlier in section 1.4, the existence of ideal measurements is significant for our motivating problem.

With these remarks I conclude my overview of "algorithms" that purport to show or make likely that empirical irresoluble alternatives exist with necessity. I cannot claim to have covered all algorithms or variants that exist in the literature, but I am confident that the same or similar reasons as the ones given in the preceding paragraphs will be effective against these as well. None of them appears to furnish a good reason to uphold a strong underdetermination thesis capable of supporting anti-realism.

Chapter 5
Problems of Representation

The object of this chapter is the idea that the origin of empirical underdetermination lies with our conceptual representation of the facts. By moving from simple to more sophisticated and radical proposals I aim to show that conventionalism(s), fictionalism, etc. are unsatisfactory responses to the problems posed by underdetermination.

5.1 Ambiguity

In the present section I examine the idea that observational underdetermination is caused by attributing to certain expressions common to the conflicting statements erroneously the same meaning or referent. Once analysis and reconstruction shows that critical expressions in those statements have actually variant meanings the conflict evaporates as one resting on an equivocation. The conflict of assertion, rather than being substantial and matter of fact, is merely verbal. Quine advocated disambiguation for resolving logical incompatibility between equivalent "systems of the world" in Quine (1992a, p. 97).

If one speaker sincerely asserts "Schliemann discovered Troy" and another "Schliemann did not discover Troy" we tend to think that one must be right and the other wrong, provided Schliemann and Troy existed. *Both* speakers could be right (or wrong) at the same time if by 'Schliemann' each refers to a different person or by the name 'Troy' to a different ancient city. The conflict of assertion could be avoided by each writing or intoning consistently "Schliemann$_1$ discovered Troy", "Schliemann$_2$ did not discover Troy", etc. instead. The example of disambiguation is trivial and inefficient, since the speakers have other independent means to discover if or if not they contradict each other. Turning to a less trivial instance, empirically irresolvable conflicting statements embedded in theories are necessarily among the statements which contain theoretical expressions non-trivially. Suppose the two conflicting statements are "all electrons have physical characteristic ϕ" and "no electron has physical characteristic ϕ", where both 'electron' and 'ϕ' refer to non-observables and so do not occur in observation sentences. Suppose further that the statements are embedded in two theories in such a way that no observation can support the one over the other however indirectly. This is the situation of empirical irresolubility. Once we "disambiguate" the two claims by renaming systematically 'electron' as 'electron$_1$' in the one and as 'electron$_2$' in the second embedding theory (say) any observational irresolubility vanishes (similarly for ϕ if necessary). Each theory makes now apparently assertions regarding different kind of entities.

No one, to my knowledge, has seriously proposed this 'solution' in this setting (but see below). And for a good reason. A non-observable object's reference is not, after all, entirely determined by the place its symbolic representative has in the systems of laws that make up the theory. Descriptivism, i.e. the view that what an object is, is fully determined by its place in the closely knit structure of explanatory hypotheses (see section 5.3), is simply wrong in this case. What an object is, what properties it has, how it can be identified, is partially determined by salient phenomena and experiments (and kinds of explanations). The reference of the term 'X-ray', for instance, is in part determined by the famous (reproducible) experiments and pictures Dr. Röntgen made. Our systems of hypotheses are, as a rule, partially continuous with independently established background information and predecessor theories. Insofar alternative theories lend itself to an explanation of the same salient phenomena, one is reasonably justified in concluding that they speak about the same entity as causing the salient effects.[1] If we legislate, to make things vivid, that Bohr made assertions about a different kind of negatively charged micro-entity than

---
[1] I focus on concrete objects to make things vivid; similar considerations apply to physical predicates.

D. Bohm, then the interpretations of atomic theory they championed conflict on no statement. The much-discussed conflict between these two empirically equivalent 'interpretations' is resolved. However, the two interpretations are actually about the very same objects, since (i) both interpretations deal with the very same ('salient') phenomena, and (ii) both ascribe the same set of characteristic physical properties to the purported entity.[2] What more can one want for identity criteria? It is fair to conclude, that the alternatives refer to the same kind of object in the world by different names. Disambiguation fails to resolve the conflict. I do not want to be dogmatic about non-descriptivism. But this consideration seems to indicate that 'disambiguation' is not the presupposition-free, metaphysically neutral procedure it appeared to be.

Does disambiguation fare better at the level of what Quine called "global theories", i.e. a sum total of all knowledge, scientifically vetted, instead of scientific theories in the usual sense?[3] The crucial difference to the previous case, from the descriptivist point of view, is that there is no independent "back-ground information" to call upon in the all-inclusive system of beliefs. Disambiguation remains a deeply unsatisfactory move. The reason is neither its triviality nor indeed that the incompatible, ambiguous alternative is still 'on the books'.[4] Disambiguation aims to demonstrate that the original conflict does not signal a flaw or deep-seated limitation in our reasoning about the world. Rather the conflict is made to appear as due to an unfortunate initial choice of words, a contingency. Although the pre-disambiguated alternative is still 'on the books' one ends up with two empirically deductively equivalent, syntactically compatible theories. In a sense,

---

[2]The notions of "salience" and similarity here are less than clear, but in concrete cases, like the ones mentioned in the text, they are easily sharpened and explicated. My non-descriptivist stance is shared, for instance, by A. Goldman, compare the footnote on p. 193.

[3]Quine attributed the idea that disambiguation can resolve logical incompatibility between equivalent "systems of the world" to D. Davidson. *Both* alternatives, after disambiguation, can be true together, in conformity with (RT), see p. 187. M. Dummett had made a similar suggestion earlier.

[4]This is a criticism voiced in Bergström (1990). It was rebutted, correctly in my mind, by Quine in the same volume.

there is no 'fact to the matter' whether there really is a logical incompatibility at the heart of our total knowledge of the world or not. Disambiguation shows logical incompatibility to be spurious and inconsequential in point of skepticism. The objection that one could reverse the procedure is beside the point.

Intuitively one expects that there is an objective fact whether or not the renamed entities in the disambiguated alternative refer to the same objects or not. This intuition has a good basis, I believe, the same one that supported the parallel case for limited domain theories. If apparently different (kinds of) objects play the same roles in causal accounts of certain salient phenomena, then that is strong reason to suppose that the objects are in fact identical. Phenomena and their causal explanations can be *referential fixpoints* for non-observational object terms. Different names can, after all, mask identical reference. It is this point that creates an obstacle for disambiguation and makes it an unsatisfactory response to empirical irresolubility at the level of 'global theories'. Disambiguation is, in effect, a Ramseyfication of the object's representative in both theories, see section 2.2. It suffers from the same defects and imports a strong element of instrumentalism into the overall account (perhaps against Quine and Davidson's intention).

5.2 Conventionalism: Local

In the previous section I have examined the suggestion that disambiguation of one or more expressions in conflicting statements will be sufficient to avoid some of the threatening relativistic or positivistic consequences of empirical irresolubility. The thesis before us now is:

> *If* two theories or systems of belief are strictly observationally irresolvable but conflict on a class of statements, then the conflict can always be traced back to different choices of key *definitions* for expressions referring to non-observables that appear essentially in the conflicting statements.

There is then, as a consequence, no 'fact to the matter' which of the two (sets of) statements is true, which is false. In a slightly misleading sense one may say that these *two systems say exactly the same thing*. Hans Reichenbach was the first to defend the thesis. Which arguments support his claim? First of all, modern theories of particle dynamics in space and time, gravity and cosmology provide what may be the most suggestive cases underpinning the thesis. I will not here review the Reichenbachian analysis of relativity theory and subsequent criticisms since this is familiar ground and there is no space here for anything like a comprehensive discussion of these matters.[5] Instead I turn to arguments of a general character that appear to support the thesis.

"Any description of the world presupposes certain postulates", Reichenbach wrote, "concerning the rules of the language used in the description. The description of unobserved facts depends on certain assumptions concerning causality [...]" (Reichenbach, 1938, p. 139). There are then propositional presuppositions, framework principles, that enable the validation of observational claims without being independently testable themselves (this is what I call a "cluster structure", see below). For instance, beside the "postulates" of causality, there are "postulates" that regulate the identification of objects across time and space, and stipulations regarding criteria of cognitive significance themselves. Reichenbach considered the (meta-philosophical) choice among meaning stipulations as unconstrained by fact and requirements of rationality, hence governed by purely pragmatic goals. Statements about the external world of common objects are not, against the claims of *hypothetico-deductive realism*, inductively supported by observation – rather their validity depends on prior conventions. The concept of *causality* is of particular importance to Reichenbach's account. Causality permeates the common notion of objecthood: objects persist unchanged in space and time when not observed (for sufficiently short periods of time); they are publicly perceptible and do not interact with

---

[5]See, for instance, Friedman (1983); and chapter 2 for a discussion of aspects of the issues. Richard Miller focuses narrowly on Reichenbach's discussion of the interplay between metric and universal forces in generating equivalent descriptions (Miller, 1987, pp. 422, 432–433) in order to refute this instance of theory underdetermination.

observers other than through purely physical interactions. The claim is now, that (a) the notion of causality is a (*de facto*) pre-supposition of all our engagement with and inductive inferences about the common sense world, and (b) there is no non-circular way of testing independently aspects of the notion of causality, i.e. the observer-independent existence of objects. Because of (b) Reichenbach suggested that the common notion of causality does not correspond to a "fact" about the world, but is a *stipulation*. (His usage of the word "postulate" here is misleading in suggesting that it is a testable hypothesis.) The contrary hypothesis – what appears to be material objects actually violate one of the causality requirements – can be made to fit with all subjective experiences, according to Reichenbach, if a-causal interactions between privileged observer and objects are postulated. To make this point, he undertook to explicitly construct the alternative "ego-centric language" and thus show the conventionality of the common notion of causality. Observationally irresolvable conflict in the area of common sense beliefs is fully analyzable into different "volitional resolutions" along these lines.[6]

Note, Reichenbach did not infer the existence of radically variant conceptual schemes from the very *theoreticity* of our experience. Theoreticity of experience comes in two kinds. One is the often-cited default conceptualization of our perception as perception of particulars in space and time. The other kind is exemplified by implicit assumptions we make regarding causal relations and identity conditions that are said to govern, if not outright define, the customary notion of bodies in space-time. One could advance an argument like this in support of the view that empirical underdetermination of our view of reality is inevitable. Is the argument valid? I think not. The inference from theoreticity of experience to multiplicity of genuinely variant conceptual schemes is a *non-sequitur*. The fact stated in the premise does not necessarily inject a perspectival element into claims about the world. Even if it does, it does not follow that other (hypothetical) perspec-

---

[6]There "are certain elements of knowledge [...] which are not governed by the idea of truth, but which are due to volitional resolutions, and though highly influencing the makeup of the whole system of knowledge, do not touch its truth-character" (Reichenbach, 1938, p. 9).

tives are not simply trivial notional variants of ours. Nothing less is required. Although the concepts we use are *our* inventions, as Einstein and Infeld remarked correctly (see section 1.1), this by itself does not show that our conceptual scheme cannot be fully representational and objective (in both senses of the word). In other words, nothing in the argument as it stands precludes that our version is a faithful version of the "absolute conception" of the world. Reichenbach's point rather is that what is theoretical or constitutive of experience is not itself grounded in experience and so is not "governed by the idea of truth".

Reichenbach originally applied the foregoing analysis to formalized scientific theories and it is to this more precise and detailed setting that I want to turn now in order to examine his approach. Most visible among the presuppositions of the Reichenbachian thesis, in the setting of proper science, is a by now classical view of the nature of scientific theory. The reconstruction of any such theory begins, according to the classical view, with the identification of primitive observation predicates and observation statements in the underlying language. The cognitive meaning of these expressions is uncontested and unproblematic. It is different with all other statements. For sentences that are neither observation sentences nor analytical sentences (i.e. logical and mathematical structures) to have cognitive meaning they must be *capable* of (direct or indirect) verification or falsification. This is the verification requirement that has already been discussed. The bridge between the purported statements about non-observables is provided by a (finite) set of special statements, the "coordinative definitions", that render a symbolic construction (the prospective theory) testable, significant and applicable. They define explicitly or implicitly the meaning of a primitive descriptive terms of the theory. *Coordinative definitions* are the logically first point of contact, in a proper reconstruction, between primitive theoretical terms and neutral (i.e. independently justified) observation sentences. (With the defining sentences a – still restricted – conventionalism enters the account.) All testable consequences of a theory are derived jointly from coordinative definitions, hypotheses and initial data (i.e. observation statements). Empirical and non-observational statements can count as objective deliverances of nature only relative to such prior stipulations. In cases where the coordination is of a concept with a physical object or with a definite, surveyable process, Reichenbach may be interpreted as explaining how *reference* of primitive descriptive terms in a physical theory is determined. The "correlations" with concrete things, the simplest case, are established by acts of pointing, ostentative definitions. An example is the coordination of "unit of length of an interval" to the Paris meter. Another well-known example of a coordinative definition is the stipulation that rigid measuring rods remain congruent when transported. If two researchers happen to ascribe conflicting values for the distance between two given points, and all errors have been eliminated, then the reason for the manifest disagreement on objective matters is that they use unwittingly different coordinative definitions as basis of their measurements. Reichenbach argued that there are *prima facie* "synthetic" statements in the *sciences*, i.e. statements about how the (non-observable) world is, that are really coordinative definitions, or consequences thereof, in disguise. This account would not only diagnose the sources of our problem, i.e. underwrite the universality thesis of empirically irresolvable conflict, in addition it would offer a neat way to dissolve the problem. The conflicting "alternatives" are trivial variants of each other.

Suppose that Reichenbach is right. Does the analysis explain and bring into one fold all cases of empirically irresolvable conflict? If not, we have at best succeeded in reducing the scope of our problem to an extent that is still unclear. There is at least one important example in the sciences which defies his analysis in terms of variant choices of coordinative definitions: interpretations of quantum mechanics.[7] Difficulties with Reichenbach's analysis have been pointed out from various perspectives and I need not rehearse the debate here (for a sampling see Feyerabend, 1958 and Bonk, 2001a). But let me highlight two issues. (a) Confirmational holism has effectively undermined the doctrine that analytic sentences (coordinative sentences are analytic) are categorically distinct from synthetic ones in that they can never be

---

[7]One who believes that *all* ("interesting") cases of empirical irresolubility are cases of definitional conventions is Paul Horwich, who disagrees on this point (see section 5.3).

revised on account of recalcitrant evidence. If coordinative definitions are no longer taken as analytic sentences, then Reichenbach's claim that there is a plenitude of "volitional choices" leading to equivalent descriptions becomes problematic. First, differences between coordinative *hypotheses* (synthetic statements now) have to be matched by changes in the original hypotheses as to leave the empirical content undisturbed. What guarantees that this is always possible, that compensating formal expressions exist? Why believe that for any statement we wish to adopt as true in the language of the theory there is a *rationally acceptable* one that exactly cancels any unwanted empirical consequences resulting from our choice?[8] Only the label "convention" could create the expectation that our common standards of units of measurements are replaceable in arbitrary ways. Only the presumed conventionalist character of those statements would explain why theoretical alternatives can be interpreted as variants of each other. Consequently, Reichenbach's analysis loses its potential for solving the problem of empirical irresolubility. The second issue (b) is that the role Reichenbach assigns to coordinative definitions is inconsistent with scientific practice, despite being based on "lessons" from relativity theory. Coordinative definitions are purportedly true by convention and can be chosen freely, altering the laws. Choice of the rubber band standard as unit of length (say), or the Dalai Lama's heartbeat as a unit of time, both entertained by Reichenbach as logically possible choices, would complicate the physical laws infinitely. Yet, in practice scientists tend to think that the form of the laws is fixed (directly by the facts and patterns of explanations) *independently* of the standards of measurement. In a sense then, it is the theory which – partially – determines what standards and methods are to be used for measurement. Moreover, typically basic standards of measurements antedate the introduction of new hypotheses by long, and are used across domains and scientific subject matter. These basic standards and methods play an analogous role to that of deductive logic in the sciences. All this supports the view that coordinative definitions cannot be chosen freely, and hence

---

[8]This objection is similar to an objection brought against a (strong) version of the Duhem–Quine thesis by A. Grünbaum (see chapter 3).

that they cannot be entirely conventional. (The residual freedom is arbitrariness of re-scaling quantities which has little epistemological significance.) Reichenbach would have replied that logical reconstruction, which abstracts from the contingencies of scientific practice, is the proper tool in the context of justification. Against this, I want to hold that a high degree of descriptive conformity to actual practice, in this case, *is* a criterion for a successful reconstruction of scientific enterprise.

I now turn to a second argument advanced to support the foregoing analysis, based on an observation on modern theories (like general relativity). The argument is more general and largely independent of an account of theories based on coordinative definitions in a strict sense. Theories of universal scope once properly reconstructed contain *clustered statements*, i.e. statements which are only testable collectively, according to Reichenbach. None of the statements which make up the cluster is testable separately and independently. That such structures should exist in our theories is a somewhat surprising discovery. Statements about the simultaneity of distant events and about the one-way velocity of light in vacuum; non-standard metrics and non-standard forces; path-ascriptions in quantum mechanics and the existence of causal anomalies are clustered, confirmationally interdependent hypotheses. The set of coordinative definitions and hypotheses, the whole theory then, appears as one big cluster in this sense too.[9] Clustered statements create a difficulty for a sentence-by-sentence verification criterion. The observational consequences of the theory do not settle the truth values for cluster statements individually. Reichenbach held that truth values can be assigned at will over a sufficient number among the statements in the cluster in a way that is consistent, renders the remaining ones testable and hence confers to them a status as "empirical claims". (Clearly, coordinative definitions have lost their central role at this point.) Empirical irresolubility disappears if the conflicting statements belong to clusters that have this peculiar interdependence structure with respect to the evidence. Note that this

---

[9]This is often overlooked, possibly by Reichenbach himself, because it is so natural to single out statements labelled "coordinative definitions".

account of irresolubility, if correct and minus the issue of coordinative definitions, does not give us much reason to believe in the universality, or necessity, of empirical irresolubility. Irresolubility appears as a conceptual possibility rooted in particular cases. Most theories do not appear to have this peculiar structure (e.g. organic chemistry, botany) and are ruled out as having potentially empirically underdetermined alternatives – as far as the 'circularity' mechanism is concerned.[10] If we look closer at Reichenbach's analysis we find a strongly anti-realist presumption: the truth value for any specific claim in the cluster can be adjusted at will. One may move from true to false by deciding on the truth or falsity of a sufficient number of other statements as long as the theory's empirical (deductive) content is preserved. What is denied is that these statements are objectively true or false as a matter of fact. Why not say instead that so many conflicting alternatives are compatible with the same set of observations and only one is the true? This way of stating the matter is certainly more consonant with good scientific practice. Perhaps Reichenbach would support his claim with the following thesis[11]:

(RT) A theory is true if all its observable consequences are true, and false if not.

The truth of an observation sentence in turn is either understood verificationistically, or neutrally taken as a primitive. The principle is another expression of verificationism (and is still upheld by Quine for "global theories" with the important qualification that observation sentences is understood naturalistically). Consequently, if two theories have exactly the same empirical content then it is impossible that they are incompatible and conflict with each other. Because otherwise, the argument presumably goes, one of them must be true and the other false, in conflict with RT. There are two possibilities to avoid this problem: (a) deny that the conflicting theoretical statements have

---

[10] This consequence conflicts with Quine's erstwhile belief that any theory, *qua* theory, is empirically (deductively) underdetermined. In more recent writings he had qualified the claim by replacing "theory" with "global theory".

[11] Compare the discussion in Glymour (1980b, pp. 14f).

a truth value, or (b) show that the conflict is spurious, i.e. depends on prior linguistic decisions. Possibility (a) has been favored by some as the proper response to the status of an "undecidable" mathematical assertions (like the continuum hypothesis), and there are passages in Reichenbach's works where he favors this *Ansatz* with regard to empirical irresolubility, compare Reichenbach (1938). The account above takes up (b). Of course, the difficulty arising from RT can be resolve in yet another way: by dropping the thesis as a false account of the truth of a theory. Reichenbach's argument by now is seen to hang on a thin thread, the correctness of a version of thesis that the cognitive meaning of non-observables is derived from their consequences for observables. I do not want to discuss the merits of this thesis further. For my purpose, it suffices to point out that the semantic reasons for adopting RT are not stronger than and do not outweigh the natural interpretation of empirical irresolubility as a case where truth and falsehood transcends the total available evidence.

Similar considerations apply to Reichenbach's analysis of common sense knowledge. Consider the once much-discussed (alleged) empirically irresolubility of the assertion that other persons have minds and its denial. Should we really believe with Reichenbach that the claim "all persons have a mind like myself" is true or false, depending only on whether we adopt conventionally certain standards of similarity? Some degree of conventionality is undeniable: we could *by fiat* change the meaning of "persons" to refer to Doric columns (say), and the claim would come out as false. This kind of *a posteriori* conventionality in a descriptive claim is obviously trivial and different in kind from the one Reichenbach argued for. Our prior judgements and intuitions in such cases are strong enough to outweigh a verificationist theory of meaning from which such consequences flow. Reichenbach's sketchy construction of an "ego-centric" language to establish the conventionality of ordinary notions of causality is in no better shape: in order for this "language" of a privileged observer to be viable at all the necessary modifications in common sense presuppositions keep accumulating. And they run deep. Reichenbach's account has no conceptual means for assessing the methodological *rationality* of one set of modifications against another in the initial theory, beyond the issue of "descriptive simplicity". In sum, the broader, conventionalist analysis does not succeed in showing that empirical irresolubility is generally a matter of verbal disagreement.

5.3 Conventionalism: Global

It may be helpful and suggestive to consider again apparent 'conflicts of assertion' that are cast in a specific syntactic form, like "Schliemann discovered Troy" and "Schliemann did not discover Troy". Speakers who make conflicting assertions like these may disagree about a fact, or they may not really disagree and just deviate over who is who and where is where. There is real disagreement provided the reference of the name of the place or the person is the same for both speakers, and only a verbal disagreement if not. The Schliemann kind of conflict involves nothing beyond observable facts and language and is obviously empirical 'resoluble'. Yet, it is suggestive for the general case of observational indistinguishability: it indicates ways for how to semantically resolve empirical irresolubility, and points to an explanation of how and why empirical irresolubility can arise at all. This explanation does not seem to require more than a common sense distinction between facts and the language we use to state these facts, or between theory and its specific formulation.

The general case requires some stage setting. Suppose that conflicting statements which give rise to empirical irresolubility always stand in a special relation to each other: there is a bijective mapping of expressions of the language into itself, that when applied to one statement outputs the other, conflicting statement, and vice versa. The mapping replaces the expression A by B (say) in all true statements and vice versa. This is an operation on the surface of the language, nothing is required with respect to the reference of the expressions under the mapping. Two possibilities: (a) all expressions in both statements keep exactly the same meaning (reference) under the mapping, or (b) they do not and the mapping "permutes" along with the words in the statements their original references and ranges in an appropriate fashion.

Given that a language (at the minimum) is characterized by vocabulary, syntax and a reference scheme the mapping in case (b) is in effect a translation of the original statement into a different, new language. The mapping (or translation) is 'appropriate' if its output in the new language is the *true* statement "Lucius is taller than Commodus" (say). The net result of the operation is a *notational variant* of the original statement. Under (b) the original conflict of assertions is best interpreted as one of equivocation. The problematic case is the first one (a): the mapping permutes the words and leaves the original meaning unchanged. A speaker of our language, we imagine, asserts the (mapped) statement and insists that the statement is about the very same non-observables. This, despite ascribing to the *B*s all and only the physical, chemical, etc. properties that the other speaker ascribes to the *A*s. There is no observation statement that can prove either of the speakers wrong: the *A*s and *B*s refer to non-observables, and all deductions of observation statements are preserved under the mapping (representations of the physical properties enter into equations and causal explanations, not the *A*s and *B*s themselves). Consequently, in contrast to (b) case (a) appears to describe a genuine empirical irresolvable conflict of assertions. Moreover, we have discovered a general 'semantic' method to generate empirically irresolvable conflicts.[12]

Suppose then that observaional indistinguishability takes this

---

[12]The idea of looking at permutation type conflicts is due to Quine (1975), who acknowledges a discussion of radical translation by B. M. Humphries, *Journal of Philosophy* 67 (1970) 167–178. Humphries' example is the interchangeability of *observation* predicates, quartz and sugar, in the foreigner's language. The standard example in the literature (Quine, 1975 and Horwich, 1982, p. 63) takes English as background language; in place of the terms *A* and *B* 'electron' and 'proton'; for the mapping a theory-wide permutation of the two; and as subject of irresolvable conflict any statement from the "atomic theory of matter". The claim that the mapping generates an empirical equivalent alternative may sound incredible at first. Suppose a particle detector is designed to be triggered if and only if an electron, in common parlance, flies close by. After the permutation, the alternative formulation says that the (same) detector is triggered if and only if a proton flies close by. The alternative formulation does not contradict the facts: the "proton" has all the physical properties we ascribe to the electron and so the detector will respond to a proton flying by. Similarly, an *H*-atom will be a stable state of an electron in the nucleus with a proton in orbit.

general form. I have already indicated that case (b) is naturally interpreted as one of where the same facts are expressed in different words, and that the conflicts of assertions are terminological in origin. Quine proposed that this is the proper response for both types of cases: alternatives generated by "permutation" mappings of terms are *notational variants* or formulations of one and the same theory. Although the overt conflict between alternatives thus interrelated there is no 'fact to the matter'. In support he cites that this criterion of individuation is in full accord with our intuitions and common sense (Quine, 1975, p. 319).[13] The criterion follows from this stronger one: theories that have the same *Ramsey sentence* are notational variants of one and the same theory.

Suppose a speaker insists that protons and electrons are two classes of particles "given" independently from and prior to any theoretical description, and furthermore that protons really have the charge and mass we attribute (wrongly) to the electron? Quine's answer is that there is nothing for this speaker to be right or wrong about, once the facts in the theory's domain are accounted for and the empirical content is verified. There are no facts about what 'electron' refers to beyond the theory and its empirical content. This out of the way, the criterion of individuation shows that we have it do with alternative formulations – notational variants – of the same body of hypotheses. The criterion, itself a convention, is adequately supported by our intuition in this kind of permutation cases of empirical irresolubility.

Quine's resolution has a strong anti-realist flavor. Perhaps one can do better and show that the "semantic contrarian" is actually *wrong* in her beliefs. The effect of the permutation is that the Bs now have all the physical properties we usually ascribe to the As, and that in all natural laws the Bs take up the place of the As. It is very natural to surmise, that if anything has all the physical properties of A, and has same causal role and syntactic position in all laws and explanations, than it is an A. Call this the *descriptive principle*. The principle coordinates the usage of a certain word with its reference by way of the theory (the set of true statements) which governs the term associated

---

[13]Quine introduced the technical term "reconstrual of predicates" in this paper.

with the word. Ideally then, and in a slogan, (a) the observational facts (the deliverances of nature) 'fix' the theory and (b) the theory is a description that, if 'tight', fixes the reference of the theoretical terms in a given language. Because of (a) the way the world actually is has an important part in determining the reference of our descriptive terms. If the world were different our words would refer to different things. Even if there is no observational difference, there is a systematic correlation. The best defense of the descriptive principle is best put as a question: what else if not our usage and beliefs should fix the reference of an expression? The upshot is that the contrarian who insists that Bs have all the physical properties and the same syntactical role than (what we would call) As, *and* maintains that B does not refer to (what we would call) As, is plainly wrong or inconsistent. The linguistic fact that we use A plus the truth of the descriptive principle guarantees the truth of $R(A, B)$ and hence the falsehood of $R(B, A)$ (where R denotes an asymmetric two-place physical predicate say) – provided, of course, the statement $R(A, B)$ is entailed by an empirically fully adequate theory. For instance, the claim that Bs are smaller than As, say, is false (if we maintain that the As are smaller than the Bs). The consequence that the contrarian's theory is wrong despite its empirical adequacy, follows from our conventional adoption of a specific language or usage (and the descriptive principle). Had we adopted the contrarian's usage from the start, the contrarian theory would have come out true – so our conventional usage (alone) does in no way "make" the underlying theory true.[14] This is a distinguishing feature of this account of strong underdetermination over Reichenbach's.

Horwich draws a principled distinction between the "verbal" and the "factual", between language and theory, that would be unaccept-

---

[14]Here I follow the analysis in (Horwich, 1982, 1986). Horwich formulates the main premise with help of the theory's Ramsey sentence: "Our basic assumption is that if our actual beliefs are represented by $S_1(G_1 \ldots G_k)$ then we have implicitly stipulated, and know *a priori*:

$$\exists \langle \phi_1 \ldots \phi_k \rangle S_1(\phi_1 \ldots \phi_k) \to S_1(G_1 \ldots G_k)"$$

S_1 denotes a set of true sentences in the language we are actually using as against the permuted S_2.

5.3. Conventionalism: Global

able to Quine. And although descriptivism as an account of reference is not without difficulties, I have no qualms with the foregoing analysis and dissolution of empirical irresolubility induced by bijective predicate mappings.[15] Conflicts of assertions can sometimes be traced to the unwitting use of variant formulations. I question, however, the scope and significance of empirical empirical irresolubility thus created. Are there good reasons to think that interpretative indeterminacies in quantum mechanics, alternative versions of space-time theories, the plethora of skeptical alternatives and a multitude of toy examples, etc. all are generated by the careless switching and mapping of certain words? Quine, for one, relegated permutative irresolubility to a segment of general empirical irresolubility. He noticed that the "Poincaré disk" furnishes an example not easily brought into the fold, since one of the two conflicting descriptions relies on a term that has no counterpart in the alternative description (Quine, 1992a, p. 97). (Poincaré looked at the plight of flatlanders equipped with measuring rods, who cannot discern by measurements whether their world is an unbounded and infinite Lobachevskian plane, or a closed Euclidean disk on which a certain temperature-field distorts the measurement rods. The critical term is the 'center' of the disk.) Judged by the space Quine dedicated to discussing "egalitarian", "tandem" and "sectarian" solutions of irresolubility he cannot possibly have thought that the segment of permutative irresolubility is the most prominent part of the spectrum. I do not wish to deny that it may be technically possible to bring the "Poincaré disk" and similar examples into the fold by adding to the postulates of these theories certain further "postulates". However, these additional postulates cannot be definitional extensions, since their supposed analyticity would raise the usual objections against this distinction in empirical theories. If the additional postulates on the other hand are "synthetic" statements, then – under

---

[15]Goldman, for instance, criticizes, correctly in my mind, the claim that "the meaning of theoretical terms accrue solely from their places in their respective theories" on the ground that physical predicates applied to theoretical objects (e.g. momentum) draw *meaning from contexts other than* the respective "framing" theories (Goldman, 1988, p. 266). These contexts are crucial experiments and paradigmatic textbook explanations of 'salient' phenomena.

the assumptions of descriptivism – their presence will in general alter the reference of all terms in the original theory. This dilemma and the obvious *ad-hoc*ness of any such extension of the original theories apart, all that could be shown by way of postulatory extensions is that *another*, different rival theory formulation exists that is related by a permutation-like mapping to one of the other formulations.

Horwich is of a different opinion. Although he does not address the example due to Poincaré, he makes a case for interpreting and resolving Cartesian skepticism along these lines. In examining this externalist argument I hope to make plausible that permutation induced irresolubility is indeed not the paradigmatic case of empirical irresolubility.[16] Let p be a common sense item of knowledge, like "Snow is white". The skeptic casts doubt on our knowledge by a constructing a skeptical alternative, for instance: "A vat-operator (the Cartesian demon) induces us to see and believe that p". Horwich suggests we rewrite the skeptical claim as $Sk(p)$, where Sk is the sentential operator: "A vat-operator (the Cartesian demon) induces us to see and believe that", or "It is an illusion that". The formal similarity to permutation cases is striking. There are two ways to interpret the skeptic's claims. (a) We can attribute to him an idiolect in which he perversely uses the operator $Sk()$ to *mean what we mean* by "It is true that". In that case the skeptic's claim evidently does not cast doubt on our beliefs. (b) We can interpret the skeptic literally. In that case, what he claims is either true or false; if false the problem is solved. Suppose then the skeptical alternative obtains and the world is such that $Sk(p)$ is true:

> [...] in a world that is correctly described in English by the skeptic's alternative theory formulation and whose inhabitants assert our sentences S_1, the terms 'London', 'grass', 'vat', etc. do not mean what we mean when we produce the very same utterances; and so there is no reason to suppose that their view is

---

[16]Horwich indicates that if there exist cases of irresolubility which cannot be dissolved in the manner described, they can be handled by applying appropriate *inductive methods* and simplicity considerations (Horwich, 1982, p. 64). I have argued above that this approach shows little promise to deliver the desired result.

> mistaken. Whichever world we occupy, our beliefs, expressed by S_1, preserve a constant truth value. (Horwich, 1982, p. 76)

Even though the world is as the skeptic claims it is, our statements p remain true due to an "automatic" adjustment of the meaning of our descriptive words to the world we are in. We have knowledge that p, as a consequence of the descriptive principle, according to Horwich, *irrespective* of how the world actually is. Thus global conventionalism removes Cartesian skepticism.

Horwich's intriguing argument is similar to Putnam's celebrated "refutation" of skepticism in the first chapter of his *Reason, Truth, and History*. It would lead us far astray to examine in full this complex externalist argument against skepticism. Waiving the worry that the skeptic's position is not correctly characterized as *asserting* this or that about the world, I briefly note that on this account we do not *know* what our words and true statements actually refer to. The brain-in-the-vat (I refer to Putnam's thought experiment) knows that 'vat' refers to vat, which is true in all worlds, yet it is unable to discern whether his usage of 'vat' refers to the vat in our or in the vat-operator's sense. This is certainly a counterintuitive consequence.[17] The proposal is not a 'solution' but a vindication of skepticism about the external world. To claim that we have knowledge, but do not know what this knowledge is about, amounts to a *Pyrrhonian* victory over the skeptic. For my purpose it suffices to emphasize another point. A precondition of Horwich's analysis is that the common sense world view and the skeptic's alternative *can* be understood as potential *notational variants* of each other. We *may* regard them as variant formulations of one set of beliefs. This is accomplished by encapsulating epistemically crucial expressions in two complex sentential operators ("It is true that"

---

[17]To illustrate: a flatlander on Poincaré's disk may utter "We live in an unbounded and infinite universe" and think and believe she lives in an unbounded universe. If her world is really a Euclidean disk, then her expression "unbounded" will refer to what we mean by "bounded and finite". She literally would not know what she is talking about. Provided, of course, the brain in the vat is able to refer at all. Putnam's 'refutation' of Cartesian skepticism may be interpreted to deny just this in an attempt at a *reductio ad absurdum*.

and $Sk()$). *Can* they possibly mean the same thing if uttered by two speakers? We understood how descriptive words A and B can mean the same thing (in different languages), because the two words played identifiable identical (syntactic, explanatory) roles in an underlying theory. The epistemic operators, on the other hand, live in the metalanguage and are not constrained by any set of first-order beliefs. Hence there is no objective reason to think the operators could have the same meaning (in different languages). In the absence of a good reason to override our firm intuitions then, we should accord these operators their usual – conflicting – meanings. Common sense beliefs and the skeptical alternative cannot possibly be seen as mere *notational variants*. Is the skeptical operator "A vat-operator (the Cartesian demon) induces us to see and believe that" perhaps self-adjusting, like common sense truths remain true across worlds due to world-correlative changes of meaning in all words? While the descriptive principle is plausible and defendable for descriptive terms and beliefs formulated using these terms, there is no independent reason to think that our second-order beliefs are thus closely coordinated with whatever world we are in. In fact, Horwich invokes not the descriptive principle, according to which the theory, a set of empirically adequate beliefs, determines the reference of a class of descriptive words. Rather, he implicitly invokes a much stronger and even more controversial externalist principle, according to which our statements are directly keyed to the world we happen to live in. I conclude that the attempt to analyze general empirical irresolubility in terms of predicate switching and permutations founders. Although there are cases that can be reasonably resolved in the way Quine or Horwich have mapped out, these form a small segment of irresolubility.

5.4 Verification and Fictionalism

"Mögliche Erfahrung ist das, was unseren Begriffen allein Realität gibt" (K.d.r.V. B 517/A 489). Kant's analysis of the four antinomies is characterized by three distinctive features. (a) The kind and number of rationally (and empirically) irresolvable conflicts is limited by the number of "cosmological ideas" derived from the four kinds of categories of pure reason (quantity, quality, relation, modality). Reason is driven to make universal claims in terms of the four *transzendente Naturbegriffe*. (b) All four anti-theses make assertions about a problematic concept which refers to an infinite totality (e.g. a limitless number of material parts, an endless chain of causes). The corresponding theses are the four 'finitary' alternatives.[18] Finally, (c), an *a priori* criterion is advanced for which concepts are "ganz leer und ohne Bedeutung" and which are not. Strawson has attributed to Kant a "principle of significance" that would turn him into a spiritual relative of modern empiricism (Strawson, 1966, p. 16). Strawson's interpretation is controversial and I will not enter into this dispute. The question, however, how such a principle of significance should be formulated and how it could contribute to the dissolution of empirical irresolubility is our exclusive concern in the present chapter.

With regard to (b), one may critique Kant for relying on an outdated concept of infinity in the analysis and statement of the first two cosmological antinomies (by Russell for one). The appeal to "infinity" in empirical (or cosmological) assertions does not need to pose a problem *per se* for us any more. That may well be so and I sympathize with this complain, but despite whatever light set-theory since Cantor has shed on this matter, a critical perspective on infinite totalities as given objects prevails, for instance, in mathematical intuitionism, finitism and certain forms of verificationist semantics. *If all* cases irresolubility could be retraced to infinitiary claims, then Kant's line of attack would still be attractive. The majority of known examples of empirically irresolvable conflicts, a few of which we have encountered in previous chapters, do not depend crucially on infinitiary claims. Kant resolved the first two antinomies, we recall, by showing that they are predications of a *"self-contradictory"* concept, i.e. 'world' understood as a thing existing in itself and as *Sinnenwelt* (we cannot meaningfully attribute, according to Kant, a limit to the world of experience, for

---

[18]I suppress the fact that Kant saw a difficulty with the notions of finiteness or limitedness as well, as when he criticised that the idea of a finite dividability of matter founders: "Denn dieses Glied lässt noch immer einen Regressus zu mehreren in ihm enthaltenen Teilen übrig" (K.d.r.V. B 515/ A 487 - B 516/ A 488).

instance). Can we generalize this idea, that empirical (rational) irresolubility is always traceable to a single crucial term which is – in a sense – self-contradictory? We lack Kant's conviction that there is only *one* such concept (see below), so one would have to check each instance of empirical irresolubility in this respect. I know of no case that can be resolved in this way. To give one example: an arrangement of physical particles can be counted in the common way or 'mereologically', and its cardinality will differ accordingly (Putnam, Goodman). There are no physical facts, it appears, that could tell us what the true cardinality of the arrangement of particles is. Mereology is a well-defined formal system and the notion of cardinality is uncontroversial. A consideration of Poincaré's disk supports this point: none of the crucial concepts ('bounded', 'center') is ill-defined. This kind of irresolubility then cannot be resolved in the way suggested by the Kantian analysis.

Closely related is Kant's assertion (a) that the kind and number of irresolvable conflicts is intrinsically limited to four pairs of theses. A finite *list* (L) as to which theoretical concepts or types of statements inevitably lead to empirical irresolubility, and which cannot not, would represent a step forward in solving the problem. A list-type description of 'problematic' concepts (a "black list") would indeed be the simplest solution to our problem. Given such a list of concepts one could simply declare them *non grata* in our system of beliefs, for the very good reason that they give rise to antinomies. With the proviso, of course, that the concepts on the list are not *indispensable* for a complete and detailed description of our world. This latter condition seems to be satisfied for the cases Kant discussed. However, the crucial step in Kant's claim (a), the derivation from the table of categories, has lost the appeal it once had. We do not recognize anymore *a priori* fixed, finite set of basic concepts for thinking about the world or which are constitutive for the possibility of experience. Carnap thought that actually *one basic concept* is sufficient to to 'constitute' whatever we may encounter in experience. Indeed, his work aimed to make plausible the claim that there is no one intrinsically right conceptual framework for our description of the world. The hopes for finding a list analogous to the one of Kant's discussion of the four antinomies of pure reason, are *nihil*.

5.4. Verification and Fictionalism

This approach then falls short of providing a satisfactory response to irresolubility: no case of observational underdetermination in the sciences, for instance, is resoluble in this fashion.

A precise rendering of (ii) has been attempted in many ways: I have recalled the Carnapian criterion on cognitive meaning in section 6.3 along with its difficulties. The difficulties of stating criteria for the cognitive meaning of individual statements in terms of supporting or infirming ranges of sense experiences (in a broad sense) motivates the exploration of an alternative approach. Here is a natural suggestion:

> If no evidence whatsoever can support or disconfirm a statement at the expense of conflicting statement, then both statements (along with their immediate consequences) have no cognitive meaning.

This is a first proposal for a sufficient criterion for the *absence* of sentence meaning. The notion of 'no evidence whatsoever' needs to be tidied up, both for the concept of evidence and the claim to universality. What counts as evidence for what? Does evidence refer to unaided observations of the layman? To the possibility of such unaided observations in remote space-time regions? Are instrumentally aided observations to be admitted? If yes, then what can be possibly said about future scientific instrumentation? Etc. Evidently there is a lacuna here, but for the moment we may sidestep most of the question by restricting attention to 'embedding theories' with equivalent empirical (deductive) content. That is, we consider statements that are embedded in a theory that allows to draw testable consequences and hence to define the usual notion of empirical content and equivalence of contents. With regard to conflicts of assertions there are two cases of special interest. (i) The problematic statements disagree on ascribing F to (a class) of objects or ascribe F to different classes of objects. (ii) One of the statements ascribes (essentially) F to (a class) of objects, while F is no term of rival embedding theories. One can plausibly hope to apply a principle of economy, or Ockham's razor to the second case but not to the first. It is always possible, however, to assimilate case (ii) to (i) by first appropriately extending the vocabulary of the rival alternative which lacks the expression F to a common language, and adding the claim 'no object instantiates F'. Assuming case (i) then, with embedding theories sharing vocabulary in the area of conflict (communality of ontology), a sufficient criterion for a *lack* of cognitive sentence meaning is:

(PV) If two statements conflict and their embedding theories have equivalent empirical hd-content (and we know that), then the statements lack cognitive meaning. Or, at least one of the terms, F, shared by conflicting statements and not appearing essentially in any observation sentence, lacks cognitive meaning.

The criterion is weaker than the full verificationist one in that it does not propose or require an explication of cognitive meaning. It states conditions under which a term, or a statement, lacks meaning – it does not say what meaning is. Thus the weak criterion escapes the difficulties full-blooded verificationism faces. The qualification in brackets is important. In each of the embedding theories, taken separately, F is an undistinguished term and as meaningful as any other. If we are not certain that the two embedding theories are observationally (deductively) equivalent, i.e. have a proof, then it will be difficult to sustain the claim that F is a concept without meaning. (Clearly, the meaning of a concept cannot generally be identified with its extension.) A *caveat*: the statement PV amalgamates two steps that perhaps should be kept separate. First, under conditions of strict and recognized empirical hd-equivalence conflicting theoretical statements lack a *truth value*. Second, descriptive statements that have no truth value have no cognitive meaning. Reichenbach, as we have seen, subscribed to both premises, but the second implication is not a self-evident truth.

Consider the following objection, based on what I call the argument from "future use in potential theories". The suspect expression F may find a role in future theories, despite the fact that both embedding theories may have been falsified. The scenario is neither implausible nor far-fetched: successor theories in the past have took over part of the vocabulary of the predecessor. This objection can be answered by clarifying that PV tells us that it is *rational* for us to take the problematic statements to lack significance, and not that they unconditionally are without cognitive significance. The formulation of PV should be emended by replacing "... then the statements lack cognitive meaning" with "... then it is rational to believe that the statements lack cognitive meaning." There is a more serious difficulty for this shortcut to cognitive significance. PV appears to run counter to good scientific practice. It would rule out statements as meaningless in cases of scientific underdetermination (e.g. quantum mechanics) where scientists regard those statements as perfectly meaningful (but false). This fact speaks strongly against PV.

We do not have to look far for a natural replacement for PV, a version that would effectively counter the latter objection. The revised criterion will not be one for meaningfulness, but a criterion for *fictionality* of a term in a theory. A fictional term has no referent, no non-empty extension, yet statements 'about' the term are taken as cognitively significant. The statements seize to be proper objects of belief as far as their relation to the objective world is concerned. Their significance is analogous to that of descriptions in a literary fictional discourse. I do not want to go into a discussion of the advantages of fictionalism versus Carnapian criteria of cognitive significance. The difference is a matter of degree in relaxing the standards of significance. The fictionalist version of PV (which I do not spell out since it is fairly obvious) is more easily brought into harmony with scientific practice in cases of empirical underdetermination than the original suggestion above. We are not forced into claiming that the scientific contrarian utters literal nonsense, when describing the world in terms of his theoretical alternative. Selective fictionalism takes 'problematic' statements at face value. Remark: it is even more in line with scientific practice to describe the status and role of 'problematic' expressions not as one of fictional entities and relations, but as denoting *idealized* elements or *Grenzbegriffe* (limit concepts) in the theoretical description. Hilbert, for one, regarded the strictly meaningless statements about the infinite in analysis and number theory as "ideal propositions" in sharp distinction to self-evident finitary propositions.

Selective fictionalism is not a viable answer to the problem of empirical irreducibility either. I have two reasons for this assessment. (a) Suppose the problematic fictionalistic expressions are (intra-theoretically) *indispensable* ("necessary") for the derivation of testable statements. How possibly is one to account for the empirical success of the theory if it depends on fictional entities and relations? We seem to be caught in a dilemma: either collapsing fictionalism back into full-blooded instrumentalism, or accounting for the mysterious effectiveness of *fictional* objects and relations in applications of the theory. If, on the other hand, those problematic fictionalistic expressions are not indispensable – indeed, empirical irresolubility in many cases shows them to be *inter*-theoretically dispensable – criterion PV justifies the elimination of those expression from the theory's vocabulary. (b) A second related reason is that a fictionalistic account of 'problematic' expressions faces the difficulty of interpreting mixed statements combining observable and theoretical expressions. If those expressions are indispensable for deriving observational or testable consequence then *mixed* statements and laws are abound in which both fictional and "real" quantities enter. How can a fictional expression enter into laws about presumably real objects, events, states and causes? This is certainly not possible if laws and hypotheses are taken to express causal relationships, so one has to adopt a Humean regularity interpretation of lawhood. Even so, lawhood is usually understood requires contrafactual dependencies: if this match were dry and were struck it would light up. The prevailing account of counterfactuals is in terms of possible world semantics. How do fictional entities or predicates behave in possible worlds? The very fictionality of those expressions seems to rule out any limits on how fictional relations and objects are in non-actual worlds. Statements like: 'If Sherlock Holmes were born in Cologne he would not have killed Moriaty', do not have seem to be capable of having an objective truth value. The only clear solution to such worries would be to avoid mixed statements in the embedding theory altogether and interpret *all* expressions referring to non-observables as fictional. This radical proposal, though, is incompatible with scientific practice of course, and is balanced if not outweighed by well-known arguments for a realist interpretation of the objects of science. Selective fictionalism is a first step on a slippery slope into anti-realism. If there is a way to avoid the slide, for instance if the problematic terms are of a special kind that never "mixes", then and only then selective

5.4. Verification and Fictionalism

fictionalism is an attractive possibility. The first Kantian antinomy, for instance, depends on a predicate like "empty time" that does not 'mix' with theoretical or observable predicates into laws that are indispensable deductive premises for deriving testable consequences. Such cases are few and far in between.

One last-ditch version of the initial suggestion – both conflicting statements lack meaning if no evidence whatsoever can make a difference – now suggests itself. We assume that there is no future or past observation whatsoever that would add or substract from the relative credibility of the two statements. The new, sufficient criterion for a *lack* of objective content is:

(V) If two statements conflict and neither direct nor indirect consequences for states of affairs, past, present or future can lend credence to either of the statements at the expense of the other (and we know that), then the statements lack objective content. Or, at least one of the terms, F, shared by conflicting statements and not appearing essentially in any observation sentence, lacks content.[19]

The provisos regarding the notion of 'no evidence whatsoever' apply here as well. Nothing of much epistemological significance depends, I believe, for example on whether instrumentally aided observations are allowed or not, whether certain physically "impossible" observation should be taken into account or not: at best such restrictions and qualifications will exclude some examples, without altering the overall epistemological question. A second proviso: the criterion does not apply to classical skeptical alternatives. Cartesian skepticism questions all beliefs held about common sense objects at once, and raises a conflict of assertion for every such statement. Clearly, applying V to fully fledged sceptical alternatives produces the absurd result of questioning the content of common sense beliefs in *toto*. This said, I want to claim

---

[19]The word "content" here should not be confused with the intensional content of thoughts in the psychological sense. I wish to use content as synonymous with the (objective) information associated with a statement, or the fact of the matter, and set it apart from the empiricist's "cognitive meaning".

intuitive appeal for V. One who has invoked this criterion is Hermann Weyl.[20] Since length measurements are comparisons with a chosen standard, a hypothetical overnight-doubling of all lengths generates an observationally indistinguishable alternative description of the natural world. Weyl rejected this alternative (like the "disk" originally due to Poincaré) on the grounds that "positing a difference" is justified if and only if it results in a "perceivable difference" (Weyl, 1948, p. 86). The same argument, based on V, would invalidate the 'possibility' that God "stops the clock" mentioned before, i.e. the natural course of events is interrupted by a universal 'freeze' of the flow of time. No observation whatsoever could support or refute this alternative to the common sense view since our consciousness along with everything else is 'frozen'.[21] One who has not invoked V is Quine in his discussion of hypothetical "global systems of the world", i.e. the sum total of scientific and common sense truths were they do not conflict. Quine held, as we have seen in section 5.1, that *logical*, syntactical conflict between global systems is inconsequential, since it originates in an unfortunate choice of notation. What if we disregard this particular argument, based on Davidson's disambiguation? There is strong textual evidence that Quine would have indeed concluded that the statements on which the competing versions clash are without meaning:

> If a question could in principle never be answered, then, one feels that language has gone wrong; language has parted its moorings [observation, or "observable circumstances of utterance" – T.B.], and the question has no meaning. (Quine, 1973a, p. 67)

I see two difficulties for proposal V. (a) The criterion for lack of content is blatantly *ad hoc*. There is no independent reason to adopt

---

[20] Peacocke defends a proposal (based on his "discrimination principle") that, despite his criticism of verificationism, is similar to (V) in Christopher (1992, p. 199). His argument is intricately interwoven with his intriguing, but controversial theory of 'possession conditions' for concepts, and cannot be examined here.

[21] *Reasons* may weigh in though, for instance, the very desirable compatibility with the special theory of relativity.

5.4. Verification and Fictionalism

the criterion except its apparent efficiency in disarming certain empirically irresolvable conflicts. Of course, one may attempt to derive the criterion from an explicit criterion of cognitive significance, or stated procedures of verification. Apart from the problem of properly grounding any such criterion or procedure in turn, our V is intended to escape the very difficulties that all previous attempts at formulating verificationist criteria face. The attempt to derive the criterion is doomed. Perhaps I have put the cart before the horse? Is it not intuitively plausible to discount the contents if statements outrun *in principle* our human ability to justify or disconfirm them on the basis of facts, however indirectly? The objection is evidently based on a verificationist requirement applied to global theories. I do not see what the emphasis on "beyond human ability" adds to its force that could balance the shortcomings of verficationism. (b) A second difficulty arises from the possibility that expressions, ruled out by V, may be intra-theoretically *indispensable* for deriving testable consequences from the theory's postulates. Indispensability and meaning holism are the reasons for why Quine favored the Davidsonian approach of disambiguation, or the "splitting" terms over "eliminating" them. (As noted a few paragraphs back, empirical irresolubility in some cases shows those expressions to be *inter*-theoretically dispensable.) Yet, statements involving relations or objects that appear without objective content – given V – cannot well form true premises for deductions. We seem to be thrown back to granting those statements an objective truth value and cognitive meaning, as in the discussion of fictionalism.

Let me sum up. Neither explicit nor weak versions of verificationism, encapsulated in the ascending arc of criteria L, VP, P, V, or SF, provide satisfactory and generally applicable means for dissolving empirically irresolvable conflicts.

Chapter 6
Underdetermination and Indeterminacy

My starting point in this chapter is this surprising claim: the assertion that the utterances of a speaker can be freely interpreted and reinterpreted so as to render the common sense thesis of fixed, intended sentence meanings baseless. The "indeterminacy of translation" between languages not only has close parallels to the underdetermination of theories, but positively rests on the existence of observationally indistinguishable but conflicting theories, or so it has been claimed.

I aim to unravel some of the threads in Quine's influential argument insofar as they cast light on the 'pragmatist' response to empirical irresolubility and the nature of natural knowledge.

6.1 Underdetermination of Translation

I propose to examine in this and the next section a paradigmatic inference from observational irresolubility to there being 'no fact to the matter', i.e. to indeterminacy. The case in question is that of alternative accounts of sentence meaning and requires some state-setting.

The well-known 'hypothesis', which ascribes consciousness and experience only to *one* individual, while the rest of humankind has a robot-like (or Zombie-like) existence without consciousness, programmed in a sophisticated way to act in all respects exactly like conscious and feeling humans would do under the same circumstances, is one early case of what I call empirical irresolubility. The "robot-theory", which denies mental states in all individuals save one, is observationally indistinguishable from the more mundane hypothesis – if overt behavior is all there is to go on.¹ Upon reflection it appears unfounded to infer from the fact that *I*, when experiencing a state of anger, get flushed, raise my voice and shake my fists, that others who show the same behavioral pattern are in the same emotional state: there is no possibility to *test* the latter inference independently. The "robot-theory" further undermines the argument from analogy (in *one* instance). Even an extension of the behavioral data basis by including publicly accessible records of brain events through tools like positron emission tomography, functional magnetic resonance imaging and electroencephalogram does not remove the observational irresolubility of the two hypotheses. The discovery that a certain kind of pain *I* experience is always associated with activity in a certain assembly of cells in the cortex cannot be used as a basis for the inference that a similar pattern of firings corresponds to similar experience of pain in another person. If the robot-theory is correct then there is neuronal activity of the right kind in the others' brain without them having any experience at all. This is a case then not only of observational but of *'neuromolecular indistinguishability'*. The "problem of other minds" (the "absent qualia" problem) has spawned a number of diverse proposals.² But no one, to my knowledge, has suggested that there is no fact to the matter which of the alternatives is true. Reichenbach came close to doing so.

A second, much-discussed instance of empirical irresolubility induced by data *restricted to behaviorial observation* arises from positing an intrinsic, irreducible quality of experience: color qualia. Someone who consistently picks out on demand, say, the blue balls from the red ones, who affirms the greenness of grass, etc. very likely has the

---

¹Of course, the singular individual who has a mind may decide on Turing's empirical criterion, mentioned earlier, that robots and zombies *have* a mind and feelings after all.

²The solutions range from Wittgenstein's notion of a "criterion" to Putnam's parallelizing the positing of other minds to positing material bodies as a hypothesis that has no serious rivals (Putnam, 1975a, pp. 342–361), chapter 17. Another inductivist is Elliot Sober who maintains that the problem of other minds is fundamentally a problem about induction and soluble on evolutionary and probabilistic arguments. Sober's proposal is unsatisfactory from a foundationalist perspective on knowledge, shared by many analytical empiricists. Here is not the place to examine its details.

same color sensations than we have and shares our understanding of color language. At least it seems rational to attribute to him such competence. Yet, it is *conceivable* that the subject actually has a *blue* experience when looking at a piece of *red* cloth, or a vivid red experience when seeing what appears to us as the blue water of a pool. The "inversion" of the visible spectrum the subject experiences, we can assume, remains hidden and undetectable on the intersubjective level (and even is hidden to herself in a variant of the thought experiment), *if* her color language is adjusted appropriately. The subject when queried labels her red experience "blue", her blue experiences as "red". Notice that the "inverted" speaker's usage is not a bizarre idiolect; she confirms fully to the public grammar of the color-language. The hypothesis attributing spectrum inversion to a given subject and the hypothesis expressing its denial are observationally irresoluble if overt behavior, including linguistic behavior, is the data basis. The case of *inverted spectra* in the color perceptions between two subjects has enjoyed considerable attention.[3] It has been interpreted either as the *reductio ad absurdum* of the idea of subjective qualia or as a *reductio* of verificationism, behaviorism, and functionalism in the philosophy of mind. Justified worries, in my mind, have been voiced by Daniel Dennett and others, with regard to the physical and functional possibilities of humans assumed in the thought experiment. Color perceptions are not an autonomous realm of the mind but are related in complex ways to feelings and emotions which may break the inversion symmetry so that a true "inversion" is in principle detectable. However that may be, considerations of physiological possibilities have not led to a consensus regarding the viability of the inverted spectrum hypothesis.

Reichenbach analyzed the case as a problem of *comparing* perceptual experiences between subjects analogous to the familiar problem of comparing lengths at different space-time points. Both remain *pseudo-problems* which disappear once the necessity of choosing a *standard of similarity*, e.g. for the comparison of perceptions between persons, is fully appreciated. One such stipulated standard of comparison is to

---

[3]See the extensive discussion and literature in David J. Chalmer's (1996), and Putnam (1981, pp. 79f., 94).

call perceptual experiences of two persons "similar" just in case the subjects perform equally well on all color-tests (Reichenbach, 1958, §27). (The very notion of a "quale" is, on this view dependent on the adoption of a further standard of comparison, this time between subjective impressions at different points in time in a person's history.) If we follow Reichenbach's discussion of inter-personal color inversion there is *no objective truth at stake* here, "absolute" in the sense of being independent from a framework decision. I have examined and reviewed the case against the conventionalist diagnosis of empirical irresolubility in chapter 5. Like Reichenbach, but without the trappings of sentence by sentence verificationism and truth by convention, Quine rejected the presumption that there is an objective, factual question underlying the empirical irresolubility of what an individual's color words refer to. Variant attributions of color perceptions and color language illustrate the claim that the *interpretation* of a person's linguistic behavior by another speaker of the same language is subject to an irreducible semantic "indeterminacy". The "riddle about seeing things upside down, or in complementary colors", wrote Quine (1969b, p. 50), "should be taken seriously and its moral applied widely". This moral is one of the considerations leading to the indeterminacy of translation thesis. But before I turn to that case, note that the claim of genuine indeterminacy is predicated on identifying an individual's *behavior* as the sole source and methodological basis for mental and semantic attributions.[4]

Although much discussed, it may not be entirely useless if I briefly review the indeterminacy of translation thesis. There are two subclaims. (1) A speaker's utterances can be variously interpreted (translated, understood), i.e. various meanings can be attributed to his verbal behavior, without doing violence (a) to any publicly observable facts about her and the linguistic community she is a member of and (b) to whatever methodological standards for proper interpretation there are. This is true whether the speaker belongs to the community of

---

[4]This is true for Reichenbach. In later writings Quine appears to have held that the indeterminacy persists if the behaviorist data basis is replaced with a physicalistic one (see below).

the individuals interpreting the utterances or not. (2) If two or more such systematic interpretations are empirical irresoluble in the way (1) requires, then there is 'no fact to the matter' which interpretation (translation, understanding) of a speaker's utterance is the correct or true one. A methodological situation like (1) is indicative of a genuine 'indeterminacy'. The indeterminacy in radical translation persists, according to Quine, when the organization of the physical universe were known in every detail along with the biology and neuroscience of the brain Quine (1969d). The indeterminacy is a case of *neuromolecular indistinguishability*. This is a further claim, which should be separated from (1). The indeterminacy thesis evidently goes beyond the usual degree of vagueness in everyday communication:

> The indeterminacy that I mean is more radical. It is that rival systems of analytical hypotheses can confirm to all speech dispositions within each of the (grammatical) languages concerned and yet dictate, in countless cases, utterly disparate translations; not mere mutual paraphrases, but translations each of which would be excluded by the other system of translation. Two such translations might even be patently contrary in truth value [...]. (Quine, 1960b, p. 73)

To appreciate fully the strength of (1) let us consider the notion of "translation" in somewhat more detail. The most suggestive setting for the thesis is that of a "radical translation", that is, the task of devising a translation between languages across steep cultural divides that separate a speaker and a translator. A translation manual is a set of hypotheses which correlate sentences (not words) in the target language with sentences in the home language subject to four *formal constraints*. (T1) An observation sentence uttered by a competent native speaker is correlated with (or mapped into) an observation sentence in the home language. There are no correlations of observation ("occasion") sentences with theoretical sentences in the home language in the translation script, because otherwise the distinction observational-theoretical (understood as gradual or sharp) would be obliterated even for the home language. Another motivation is the following consideration. The utterance of an observation sentence by a speaker in the presence of the observer is thought to be directly stimulated by a concurring signal event in their shared environment. The translating observer will similarly correlate the concurrent event (the stimulation caused) with an observation sentence in the home language. (T2) Conjunction, disjunction and negation of sentences in the target language are identifiable (by trial and error) and correlated to conjunctions, disjunctions and negations of sentences in the home language (the correlation function or recursive mapping "preserves" truth-functional compounds). In fact, the native's *syntax*, or grammar, is discernable and is thus not subject to the indeterminacy of translation. (T3) Stimulus-analytic and -contradictory sentences in the target language are empirically identifiable; the native speaker assents to them, when queried, for every class of global stimuli. The translating observer correlates stimulus-analytic or -contradictory sentences with corresponding sentences in the home language. (T4) The observer can discern what sentences are synonymous for speakers of the target language on empirical grounds (Quine, 1960b, p. 68). The potential translations of a sentence taken, say, from the native's advanced theoretical sciences or from the native's theology are not significantly restricted by requirements (T1) to (T4). Nor is the further partition of sentences into word-like expressions (nouns, particles, etc.) determined. Quine called translation hypotheses in these cases "analytical hypotheses" because they appear to be uncheckable *independently* from the success or failure of the translation, given nothing but observations of overt behavior of the native speaker. (The identification of the speaker's assent (to a yes-or-no query) as "assent" and the speaker's dissent as "dissent" is an analytical hypothesis.)

One may want to add further requirements to the four above in order to narrow down translations to the point of uniqueness. Why should certain kinds of hypotheses which have been successful in the translation of *other* alien grammatical languages, for instance, or the discovery of cultural universals, not be used in shaping the translation manual? In the sciences theory formation is frequently guided by collateral information from back-ground theories; I discuss the view that those 'transplanted' hypotheses carry confirmational clout in more detail in chapter 3. Quine did not object to extending the list with additional clauses. But since such hypotheses are "analytic" hypotheses their imposition on the formulation of the present manual is strictly a matter of *decision* or conventional choice (volition). Quine believed that conditions (T1) to (T4) *exhaust* the contribution of observation to the creation of a "correct" translation manual (and has been criticized for undue "essentialism" on this point).

Consider two *empirically indistinguishable* translation manuals of French into German: the first correlates the observation predicate *chien* with "Hund(x)", the second correlates *chien* with "Hund(x) ∨ Einhorn(x)". The method of pointing to various animals in the presence of the native French speaker, uttering "chien?" and registering assent or dissent cannot discriminate between the two translation manuals since unicorns do not exist. Of course, the translator can confront the speaker with, for example, a *pictographic* representation of unicorn. (I doubt, however, that Quine can allow the speaker's response to a picture (in lieu of a real exemplar) as evidence.) A translation dictionary which correlates *chien* with the (non-observational) predicate "Hund(x) ∧ '3 is a prime number'" in German shows that an enlargement of the test and data basis would not necessarily remove the irresolubility (Carnap, 1955, p. 238). (To bring the example in line with the recent literature consider replacing "chien" by "Gavagai", "Hund" by "rabbit" and "Hund or unicorn" by "undetached rabbit part".) A more sweeping example is furnished by the following pair of empirically irresoluble *English to English* translation manuals.[5] Manual 1 "maps" every sentence into itself and interprets a speaker's assent (dissent) to a query as assent (dissent). Call it the identity translation. Manual 2 "maps" every sentence in its negation and interprets a speaker's assent

---

[5] The scenario of a "radical translation" was meant by Quine to drive home another lesson: indeterminacy besets the interpretation of utterances between speakers of the *same* language. This is what the color inversion case is supposed to illustrate. "Thinking in terms of radical translation of exotic languages has helped make factors vivid, but the *main* lesson to be derived concerns the empirical slack in our own beliefs. For our own views could be revised into those attributed to the compatriot in the impractical joke imagined; no conflicts with experience could ever supervene [...]" (Quine, 1960b, p. 78; my emphasis).

as dissent and vice versa. The manual satisfies all four requirements above. A related, more sophisticated manual 3 is readily constructed for a first-order language with identity and without singular terms exploiting the dualism between conjunction and disjunction, existential and universal quantifier, necessity and possibility in first-order (modal) logic (Massey, 1978, pp. 50–52; Putnam, 1975b).

Claims (1) and (2) (above p. 6.1) have a controversial consequence: what a speaker "means" by an utterance and what her words refer to is not determined unconditionally or absolutely given by what goes on in her head or by the rules of her language community. Rather, it is relative to a translation manual chosen by the translator. These claims are at odds with two accounts of sentence meaning (not to say with common sense). A proposition is, according to the classical theory, a linguistically neutral, abstract object, which is grasped by the mind, when it understands a sentence. Russell for one maintained such a conception of sentence meaning: "Does logic or theory of knowledge need 'propositions' as well as sentences? Here we may define a 'proposition', heuristically, as 'what a sentence signifies.' [...] we can make sure of some meaning for the word 'proposition' by saying that, if we find no other meaning for it, it shall mean 'The class of all sentences having the same significance as a given sentence'" (Russell, 1921, pp. 208, 209). To every declarative sentence corresponds a definite separate proposition in the mind of the speaker. Propositions are, on the theory that Quine attacks, regarded as the *content* of sentences in the language of the thinker or of a *thought*, which is usually a belief "that p", e.g. that a is F. The apparent uniqueness of translation between different, sufficiently rich languages has a natural explanation in that it is the same idea or proposition that is differently expressed in different languages. (1) and (2) are also at odd with Carnap's empiricist account of meaning, who is not guilty of hypostatization of abstract entities. Among those who understand an individual's ability to use a language as her having an interconnected system of dispositions for verbal responses Carnap believed that extension *and* intension of the speaker's expressions can be empirically determined (up to the usual inductive uncertainties). For instance, an observation predicate's intension, or meaning, is understood as the class of *possible* objects (or logically possible cases) which satisfy the predicate. "[...] the assignment of an intension is an empirical hypothesis which, like any other hypothesis in linguistics, can be tested by observations of language behavior" (Carnap, 1955, p. 237).[6] In the process of *empirical* meaning analysis linguists can narrow down (eliminate alternatives) and eventually discover what intension the speaker assigns to, say, an observation predicate like horse by systematically querying the speaker of a foreign language. If translation is indeterminate then the Carnapian project is fundamentally misguided, since the meaning of a linguistic expression is no more a proper object of empirical inquiry than a physical state of absolute rest, say.

The question before us now is what are the requirements for cases of empirical irresolubility to indicate an indeterminacy in the question at stake? Indeterminacy would not only resolve the empirical irresolubility, but also explain its existence. We are looking in particular for reasons to adopt claim (2) from the beginning of the section.

6.2 Indeterminacy versus Underdetermination

In arguing for the indeterminacy thesis Quine did not rely on illustrating examples. The main *systematic* reasons for the thesis have to do with the (strong) underdetermination of scientific theory by observation. I examine two reasons. (1) There is the obvious strong *analogy* between the observational irresolubility of variant interpretations of the utterances of a speaker on the one hand, and the (hd-) underdetermination of systems of scientific posits by observation and tests on the other.[7] Anyone who grants the *underdetermination* of "global science" should be inclined to embrace the indeterminacy thesis (that is accept claim (1) in section 6.1) with regard to translation as well: the methodology is similar in all respects. (2) If the thesis

---

[6]Carnap referred to an early empirical study of semantics by Arne Naess: *Interpretation and preciseness: A contribution to the theory of communication* (1953).

[7]I drop the prefix "hd" on concepts like empirical content from here on throughout this chapter. Still, these concepts are to be relativized to the hypothetico-deductive (hd) method, the kernel of scientific method for Quine.

of general underdetermination is true in a substantial sense then that fact can be taken as a premise in establishing the indeterminacy thesis. In short, translation maps observation sentences into observation sentences, and if the physical theory held by native speaker and translator agree on all observation sentences, there is little left to constrain and exclude arbitrary reinterpretations of the native speaker's theoretical beliefs. The interrelationship between the two theses is highlighted by the possibility to make the *converse argument*: the derivation of the underdetermination thesis from the indeterminacy of translation ("right translations can sharply diverge"). Any theory, interwoven into the speaker's language, is mapped into an observationally equivalent "copy" of itself by first forward translating the speaker's sentence by one manual and "back" translating by another manual. The converse argument carries little force, however. The indeterminacy thesis is much less certain and more problematic than the underdetermination thesis so that nothing is won by deriving the one from the other (Dummett, 1973, p. 404). Still, the potential mutual derivability is a sign of how close the two theses are related.

I begin my examination with the "second intension" argument (2) for the indeterminacy thesis, often taken to be central to the claim.[8] Suppose sufficiently rich theories of physics always have indistinguishable, conflicting alternatives. Suppose the circumstances indicate that the speaker of the target language makes accurate, successful predictions based on a sophisticated physical theory. Subject to the empirical strictures on any translation of the speaker's utterances, can the translator determine which of the various supposedly underdetermined versions of the physical theory the speaker believes in? The answer appears to be No. The only avenue to the speaker-physicist's theory, a set of "analytical" hypotheses, is through observation sentences. And the underdetermined rivals share the same empirical content. No conceivable observation could, it appears, falsify the interpretation of the speaker's utterances as made from the standpoint of indistinguishable

---

[8]Dummett takes the "second intension" argument to be Quine's "one definite argument" (Dummett, 1973, p. 407). M. Solomon agrees (Solomon, 1989, p. 122). Putnam writes that the argument "strongly moves Quine" (Putnam, 1975b, pp. 179, 183; 1989, p. 229).

rival theory A, say. The translator cannot refute or confirm an interpretation (a translation manual) that attributes to the speaker-physicist, on the basis of her verbal behavior, assent to the Euclidean theory of space and time (assuming a Reichenbachian view of physical geometry for the moment).

> [...] the same old empirical slack, the old indeterminacy between physical theories, recurs in second intension. Insofar as the truth of a physical theory is underdetermined by observables, the translation of the foreigner's physical theory is underdetermined by the translation of his observation sentences.[9] (Quine, 1970, p. 179)

Quine's aim, to repeat, is to demonstrate that there is "no fact to the matter" what the speaker-physicist believes and what not. Only *relative* to further substantial theoretical assumptions on the part of the translator does this question have an answer, i.e. an empirically checkable answer. (The argument is supposed to cover the case where speaker and translator-observer communicate in the *same* language: understanding each other's utterances presupposes translation, the correctness of which is underdetermined by conditions (T1) to (T4), see section 6.1.)

The "second intension" argument is not a compelling reason for the indeterminacy thesis (for accepting part (1) of the thesis).[10] I have argued in chapter 1 that it is far from clear that the "universality" thesis of empirical irresolubility is true, i.e. that there are underdetermined rivals to any given, sophisticated theory and all theoretical statements. Granted one or two general theoretical principles, mathematical and logical considerations in general will suffice to rule out all but one alternative in a space of theories. It is difficult to maintain that a native

---

[9]Similarly: "To the same degree that the radical translation of sentences is under-determined by the totality of dispositions to verbal behavior, our own theories and beliefs are in general under-determined by the totality of possible sensory evidence time without end" (Quine, 1960b, p. 78).

[10]This is not the place to review and examine general objections, like Timothy Williamson's (1994), against the existence of (referential) indeterminacy. I found Field (2000, pp. 283–284, 293–303) a persuasive reply to the Williamson problem.

speaker-physicist rejects natural truisms on the level of, say, "energy is conserved in closed systems". There is then not enough diversity in the lot of observationally equivalent rivals to call the existence of determinate sentence meanings, or intensions into doubt. The 'second intension' argument presumes an unbounded number of underdetermined rivals. If there were only a *handful* possible "genuine" rival to every theory than we are well advised to be content with the common sense view that there is a fact of the matter of which is true. Indeterminacy would give way to ignorance and uncertainty. Indeterminacy hence presumes an *amorphousness* in our theoretical accounts of the totality of experiences. The range of alternatives have to form a 'spectrum'.[11] A second reason for taking the "second intension" argument with a grain of salt is its application to translations between speakers of the same language. Suppose two physicists disagree relative to the homophonic translation manual over which of two underdetermined alternative theories (A or B) is the true one. Are we invited to believe that there exists an alternative, acceptable translation manual in which physicist 1 finds physicist 2 to *agree* with him on theory A, say? That would be an interesting way of resolving scientific controversy.

There is a plausible *objection* against an inference from empirical irresolubility in the sciences to indeterminacy thesis that must be dismissed. For the native speaker-physicist's theory to be translatable in a unique fashion into the translators language certain conditions have to be in place. If the underdetermined rival theories (in the translator's home language) conflict they do so over truth-values assigned to non-observable sentences. These sentences or laws depend irreducibly on a specific theoretical term τ, say. An *a priori* condition for the possibility of a unique "best" translation is that (a) the native speaker-physicist's theory contains a non-observable term τ'; (b) the cluster of sentences or physical hypotheses which govern τ' can be translated into the cluster of laws that govern τ in underdetermined rival theory 1 *or* 2 or n.

---

[11] This point is not universally appreciated in the literature. Stroud, for instance, defined: "The translation of a foreign sentence S_1 is indeterminate if and only if there are at least two non-equivalent sentences S_2 and S_3 in the linguist's language such that no dispositions to verbal behavior on the part of the natives are sufficient to decide between S_2 and S_3 as the translation of S_1" (Stroud, 1968–1969, p. 85).

Quite possibly the structure of the native's theory prevents the satisfaction of the conditions. In that case there is no unique translation manual. Should the indeterminacy thesis amount to the banality that sometimes conditions (a) and (b) cannot be satisfied (Dummett, 1973, pp. 405–407)? The objection misses the point: the indeterminacy thesis claims that all translation manuals which satisfy the four empirical constraints (T1) to (T4) are epistemologically *on par* (pragmatic choicess of the field-linguist aside). The "best" manual in the sense of conditions (a) and (b), in the rare case where one exists, is no truer than any other set of translation hypotheses. Inclusion of the two conditions as additional clauses with the four requirements, which define admissible translations, would indeed amount to what Quine describes as "imposing" our views on the speaker-physicist's space of conceptual possibilities.

I now turn the *argument from analogy* (1). A consideration of the "nature of possible data and methods" offers a broader and deeper argument for the indeterminacy thesis (Quine, 1960b, p. 72). The underdetermination of "global science" is, if we follow Quine, caused by the need to use concepts referring to non-observables, hence by deductivism (which gives rise to confirmational holism) with stimulus-observation sentences as potential falsifiers or confirmers. Similarly, indeterminacy in *radical* translation arises if the translator has nothing to go on but agreement or disagreement with directly observed behavior on the part of the native speaker, verbal and non-verbal alike (see below for qualifications). The similarity is striking on the *methodological level*: the trial and error method in devising translations for more "complex" sentences about non-observables in the target language, the linguists' quasi-Duhemian procedures, and the use of simplicity considerations.

> [W]hen we reflect on the limits of possible data for radical translation, the indeterminacy is not to be doubted. (Quine, 1987, p. 9)

In both cases, the criteria for correctness cannot be improved. Consequently, anyone who grants the underdetermination thesis should be inclined to embrace, first, the underdetermination of translation manuals, and, secondly, the indeterminacy thesis as well. Did Quine propose to take underdetermination as a *sufficient criterion* for indeterminacy, or is the second claim based on an additional premise? The second claim appears to depend on further premises like (i) cognitive sentence meanings, propositions, etc. do not causally *explain* behavior (are dispensable); or (ii) *ideal* knowledge of all relevant neurophysiological facts about a speaker and her environment cannot settle empirical irresolubility with respect to semantical notions. It is the inference to indeterminacy that interests us in the present section, i.e. the question why one should infer indeterminacy with respect to semantic notions like cognitive meaning, but stop short of inferring indeterminacy with respect to the posits of the natural sciences.

Chomsky was the first, I believe, to turn the analogy with empirical irresolubility on its head. He argued not against (1) but against the inference to *indeterminacy* encapsulated by (2) (section 6.1). The indeterminacy thesis, Chomsky maintained, is just another illustration of the familiar phenomenon of *underdetermination*. And since we tend to think there is a fact to the matter which of two empirically equivalent, "genuine" rival accounts of a domain of facts is the correct one, it would require additional reasons, that are not given, to claim an exception in the case of radical translation. Since empirical irresolubility does not and should not stop the scientist from positing theoretical entities, so translational underdetermination by verbal behavior (neither surprising nor threatening) should not stop us from positing sentence meaning, intentional states, etc. Yes, there are potentially equivalent rival manuals of translation, i.e. (1) is granted, but this exposes at worst our ignorance or lack of certainty about the native's speakers beliefs and meanings (Chomsky, 1980, pp. 15–16, 258, n. 33; 1975, pp. 182f).[12] Quine rejected this argument:

> [...] the indeterminacy of translation is not just

---

[12] Compare Chomsky's contribution to the volume *Words and Objections* (Davidson and Hintikka, (1969) followed by Quine's reply. Chomsky restates his criticism in Chomsky (1975, pp. 182, 249, n. 30 – n. 35). *Both* accept as true general empirical irresolubility. A full recent evaluation of the debate between Quine and Chomsky is given in George (1989).

inherited as a special case of the under-determination of our theory of nature. It is parallel but additional. (Quine, 1969d, p. 303)

'Parallel' refers to the methodological similarities between the two cases, while 'additional' indicates the persistence of underdetermination of semantic notions if there were no empirical irresolubility in the natural sciences, and all physical facts were ideally known.

Chomsky refers somewhat uncritically to the "underdetermination of theory by evidence in the natural sciences" – as if the thesis were unproblematic and plenty of uncontested examples were at hand. His reasons for accepting the thesis are a mixture of suggestive examples in the history of the sciences and Humean doubts (paralleling those of Quine) about the prowess of "scientific methodology" to resolve underdetermination of theories in the natural sciences (Chomsky, 1980, p. 17).[13] The considerations of chapter 1 have shown that the matter is not quite as clear-cut as Chomsky presumed. Let me waive this worry and turn to the alleged parallelism of scientific theory and translation manual in point of empirical equivalences. I say 'alleged' since there are stark differences, on the surface at least, with regard to (a) the kind of non-observables undermined in each case, and (b) the notion of evidence. It is a further question how the differences may affect the indeterminacy claim.

Ad (a) As systems of *hypotheses* translation manuals and "global science", or individual theories in the ordinary sense, have little in common. Obviously, the (exact) sciences aim at formulating natural *laws*. The same cannot be said of a translation manual (in the radical translation scenario): even if the native speaker's utterance "gavagai" is to be translated *determinately* and uniquely into "Hase" or rabbit, this correlation cannot count as a law. The correlation does not have the form of a natural law, there cannot be any claim to necessity, in the sense that there is a meaningful counterfactual to be satisfied. In addition, scientific theories (today) are frequently axiomatized; they

---

[13]Putnam agreed in that indeterminacy of translation (of a speaker's semantical apparatus) is similar to the underdetermination of scientific theory by evidence (Putnam, 1989, p. 229).

aim at a few, general (universal) principles; they aim at quantitative and precise results; they can be conjoined to produce novel predictions in new domains of application. A translation manual has none of these characteristics which are relevant to the question of confirmation. The parallel is incomplete, moreover, with respect to what is or is not inferred from the data. Under dispute here are not epistemic attitudes, which may indeed be taken "instrumentalistically" for the purpose of explanation and prediction of behavior, but (1) (abstract) objects of such attitudes: propositions, ideas, sentence meanings, thoughts, etc. (to which one may want to ascribe truth); (2) "categorical" objects of modern (psycho-)linguistics, like phrase boundaries, inborn conceptual categories, rules for forming regular past tense forms, etc.; and (3) psychological objects and mechanisms like the Freudian concept "repression". How plausible is it to apply the *posit-hypothesis-test* model to mind and language (the intensional apparatus), and treat them as *theoretical posits* on par with electrons in micro-physics and black holes in cosmology? The latter figure in *laws* and explanation, and are *indispensable* in accounting for the phenomena in mature theories. If we follow Chomsky, cognitive structures and linguistic objects of type (2), as mediators between an individual's experience and behavior, satisfy this criterion: they are scientifically respectable.[14] Objects and processes of type (3) are perhaps methodologically suspect, but ideally they satisfy the criterion as well. It is different with entities of (1). They can count as fallible "posits" and as "theoretical" only with respect to *other* persons. That I have thoughts and that my sentences have determinate meaning is not something I am free to entertain as a fallible hypothesis in any interesting way. The existence, moreover, of causal laws connecting items of (1) with action is contentious. The model of posit-hypothesis-test then has limits when applied to the mental. These limits indicate the greatly reduced explanatory value of this kind of mentalistic (or abstract linguistic) posit. Recognition of these limits is among the reasons for why Quine held the underdetermination

---

[14]This, not the innateness issue, is the point of his debate with Quine: the latter denied that linguistic theory, as a theory of causation of behavior, is a proper natural science (George, 1986, p. 490).

6.2. Indeterminacy versus Underdetermination

of translation manuals gives rise to indeterminacy.

I turn to the second point of comparison between underdetermination in the sciences and of radical translation, the notion of evidence. *Ad* (b) The contentious issue here is the exclusive role of observation of overt linguistic behavior on the part of the native speaker as the sole source and test basis for the attributions of specific cognitive meaning to utterances. The issue is crucial since empirical irresolubility, and hence presumably indeterminacy, comes cheap if the test basis is arbitrarily impoverished. To establish indeterminacy no potential source of knowledge is to be excluded. Potential categories of sources, beside (i) observation of linguistic behavior on the part of a speaker (extension, recording assent and dissent, etc.), include (ii) introspection, understanding, empathy, "putting yourself into the other's shoes" and similar introspective data, (iii) historical and sociological facts about the speaker's community and similar ones elsewhere, (iv) anthropological, evolutionary, and biological facts about the speaker, and (v) physical and neurophysiological facts about a speaker's cognitive processes and her environment. Cross-inferences between evidence from sources (ii) to (iv) are a potentially powerful tool for purpose of radical translation. We have already seen that background information in some cases *can* help to resolve empirical irresolubility in the case of scientific theories. It appears that Quine rejected cross-inferences of this sort in translation, whose pragmatic necessity he granted of course. Before I consider his reason I want to single out two theses with respect to the sources of information that are crucial for the indeterminacy claim. The first thesis is explicitly behavioristic:

(BR)   Agreement or disagreement with overt (verbal and non-verbal) behavior of the native speaker is the exclusive and maximal basis for judging the correctness of a translation manual that satisfies (T1) to (T4).

The adoption of certain additional rules, i.e. a principle of charity, in fashioning a translation is compatible with (BR); the thesis aims at delineating the *factual* basis for assessments of correctness. We can leave it open whether (methodological) behaviorism played the significant role in Quine's thought it is said to have, or not.[15] He conceived language as "the complex of present dispositions to verbal behavior, in which speakers of the same language have perforce come to resemble one another" (Quine, 1960b, p. 27), where the dispositions are acquired by conditioned response. Behaviorism has been amply criticized and there is no need to rehearse the arguments here, but notice that *if* language is essentially a social construct then observation of a speaker's verbal and non-verbal behavior is of course essential.[16] Consider next the *physicalistic analogue* of (BR). The thesis claims the methodological primacy of the relevant neurophysiological, etc. facts about the native speaker in assessing a speaker's attitudes and sentence meanings, and hence the correctness of a translation manual:

(PH) The exclusive and ultimate data basis on which to assess the correctness of a translation manual that satisfies (T1) to (T4) is the ideal totality of physical and neurophysiological facts (or, the best and ideally unique systematization thereof) about a native speaker's cognitive processes and her environment.

There is, of course, some reason to expect that aspects of behavior can be explained at the level of physiology (molecular biology, genetics). Physicalists claim, that if the totality of physical facts, the most

---

[15] Quine accepted a limited degree of innateness in our language faculties: natural kind concepts like 'cat' are "rooted in innate predisposition and cultural tradition" (Quine, 1992b, p. 6) He rejected the possibility (along supporting experiments) that the object category could be innate. For an account of Quine's subtle views of behaviorism compare Gibson (1982, pp. 195, 204). Gilbert Harman argued that Quine's thesis can be dissociated from behaviorism, see Davidson and Hintikka (1969, pp. 14–26). Gibson, after an extensive review of the debate, comes firmly down on the side of the behaviorism-indeterminacy connection: for Quine the indeterminacy thesis is *"logically wedded* to his behaviorism" (Gibson, 1988, p. 129; his emphasis).

[16] This is exactly Quine's view: "Language is a social art which we all acquire on the evidence solely of other people's overt behavior under publicly recognizable circumstances" Quine (1969b). The conventional rules governing the attribution of a belief, a thought, a feelings, etc., to an individual depend on the social practices and norms of the speech community to which the speaker belongs. Therefore they are interest-relative (subjective) in a manner in which hypothetical entities in the natural sciences are not (compare Putnam, 1978 pp. 41f).

6.2. Indeterminacy versus Underdetermination

complete knowledge imaginable, is insufficient to individuate mental processes, linguistic objects, etc. (as causes of behavior, or intermediate cognitive states) then nothing can.[17] One way physics or neurophysiology may be thought to ideally "determine" or individuate mental states, reference or meanings, is by identification of a set of physical states that are sufficient and necessary for a given mental state, for the individual to "mean" this and not that. The difficulties facing this strong, reductionist conception of determination by identification have led many physicalists to subscribe to a supervenience account of the relation between mental states and neurophysical states, which we need not spell out here. My point is that (PH) and (BR) function as substantive premises that are required to turn cases of empirical irresolubility into instances of "microphysical" or "observational" indeterminacy. I return to this issue in section 6.4.

In order to examine the role of systematic observations of verbal behavior and the status of auxiliary hypotheses in translation I propose to look at a similar task: *deciphering a code* or an ancient language. Ancient writing systems like Linear B, Maya or ancient Egyptian hieroglyphs, of which only fragments exist, have been successfully translated – without the benefit of interviews with native speakers, shared stimulation and observation sentences, signs of assent and dissent, knowledge of the social context, etc. The task of rendering a writing system which was not in use for many centuries readable is surely a close approximation to the radical translation scenario. True, in difficult cases a decipherment may yield nothing more than the translation of a number of ideograms or words (names, numbers, natural kind terms, titles, measures, etc.) and thus does not look much like a full translation manual for a potentially infinite variety of sentences. But surely, this is a matter of degree and not the basis of a categorical distinction between a process called decipherment and a second process called translation. The success in translating written records of societies of the remote past is impressive: linguists proceed on much sparser evidence than

---

[17]Quine concurs: "To expect a distinctive physical mechanism behind every genuinely distinct mental state is one thing; to expect a distinctive mechanism for every purported distinction that can be phrased in traditional mentalistic language is another" (Quine, 1970, p. 180).

Quine allows for and the final, full translation manual is regarded as *the* unique solution to the problem up to the usual inductive uncertainties. Part of the explanation for the success in deciphering the meaning of extant fragments are experiences with previous translations. The researcher comes to expect that the subject matter of the texts are recordings of victories, genealogical sequences of kings, calendars, laws or inventory lists, etc. She may find additional clues: a name may occur in fragments of texts from different languages, or a few signs correspond to signs of an already deciphered system. Tentative assignments of specific meanings to specific symbols (some with strictly symbolic, some with phonetic value), adducted from clues like these, are then tested against known fragments of texts and newly discovered inscriptions. The probability that a translation is accidentally successful on the extant number of fragments, if their number is large and varied enough, is vanishingly small.[18] (The "meaning" of the text thus discovered is the social, public meaning of the ancient culture under investigation.)

My brief and schematic description of the translation of a dead written language illustrates that the multiplicity of translation manuals is quickly reduced once the translator is allowed to employ auxiliary assumptions about the structure and content of the fragments.[19] The hypotheses that the round stone with inscriptions found in the rain forest is a calendar is an inference from independent, past experience. That the bow-hunting native has a bow-hunting vocabulary and know-how that matches ours is a *justified* hypothesis – even though the hypothesis is not justified considering merely overt verbal and non-verbal behavior, facial expressions, etc. In cases where a language has

---

[18] Compare Descartes' farsighted remark on this subject (section 1.1).

[19] Wallace compares Quine's account of radical translation (interpretation) with the historic decipherment of Linear B, and finds, not surprisingly, that it by and large fails to conform to the actual practice of decipherment (Wallace, 1979b, p. 230). Wallace believes that Quine's account conflicts in various ways with the specificity and variability of non-holistic "contextual frames" in which uninterpreted language fragments are presented to the decipherer. I disagree on the diagnosis and the philosophical significance he attaches to the "failure" to conform to practice (see below and Bruce Vermazen's critique in LePore, 1986).

only a small number of adjectives – the Nigerian language "Igbo" has eight – it is *predictable* that they come in basic contrasting pairs: big – small, good – bad, hard – soft, etc. (Lakoff, 1987, p. 290). A translation is constrained in yet another way: it must respect the native's rules of *inductive inference* besides those of predicate logic. For instance, induction by enumeration, a common and risky heuristic, would become practically impossible when based not on simple elementary predicates but on complex truth-functional compounds. The induction from 'object a_n is black or square' and 'a_n is a raven or a bigfoot' to 'all things that are raven or bigfoot are black or square' is less reliable than an inference based directly on elementary predicates like 'is black'. It cannot be ruled out *in advance* that a manual needs to postulate non-standard rules of induction to account for observations on the native's behavior. However, inference rules can be compared objectively in point of reliability and this allows for an objective ranking of translation manuals. Finally, biological constraints on language use need to be taken into full consideration. It is well known that the acquisition of complex predicates like "is red-or-not-round-and-dangerous" is difficult since it takes many more instances to learn. The decision whether such a compound predicate is applicable to an object or not takes a relatively longer time. Limits on memory and computational resources may rule out certain types of predicates as possible synonyms in a translation.[20] On the basis of these facts the rejection of translation

---

[20]There is solid experimental evidence that infants at an early age, before they have reached that stage in language development which according to Quine corresponds to the "reification" of things, have a concept of a thing as a spatial-temporal continuant, as something that persists when not looked at. Experiments indicate that children "know" rudimentary laws of motions, the kinematics of things. (An impressive array of experimental results is presented in Spelke et al. (1992), Clark (1993) and Lakoff (1987). A genetically anchored, non-linguistic thing-category certainly would support a notion of linguistic reference for singular terms and predicates in everyday discourse. Quine has brushed off those experiments with the remark that the children's attention span is too short: the attentive, searching gaze for the vanishing ball behind the screen wanders off too quickly to be significant (Quine, 1990, p. 7). What matters here, however, is the length of the child's attention span not absolutely, but relative to the typical attention span of its developmental stage. On this scale the behavior is exceptional and marked. At

manual which assign to terms in a native speaker's sentences monstrous chains of conjunctions or disjunctions of independent primitive terms is rationally justified. I do not overlook the possibility that the native speaker may have a different set of primitive terms (predicates) such that *our* predicates appear to her as long truth-functional compounds. But then her translation manual for the translator's language is similarly constrained. Such knowledge restricts the circle of admissible translations up to the point of uniqueness. Now Quine, as I have pointed out, granted all this:

> What the indeterminacy thesis is meant to bring out is that the radical translator is bound to impose fully as much as he discovers.[21] (Quine, 1990, p. 5)

The claim here is apparently that the auxiliaries used by the translator to narrow down competing manuals are *arbitrary* from an "epistemological" point of view, i.e. with regard to their lacking *independent* grounding in facts. The auxiliary assumptions are only loosely connected to the direct evidence (the physical inscription, the actual observation, etc.), and their transfer from one case to the next is based on

---

any rate, Quine's objection does no justice to the richness of experiments in this field. He seems to hold the view that the human mind is a "general purpose" device that is programmed by parents and various cultural influences only after birth, based on a few genetic predispositions. The researchers I mentioned understand the mind as pre-programmed (to various degrees), made up of specialized mechanisms ("modules") that predispose humans to think and act in a particular manner when it comes to basic needs like selecting a mate or fighting off sexual competitors. Others, like Steven Pinker recently, extend these mechanism to include higher cognitive functions as well. The genes for these complex mental functions, in their view, were passed on through the generations because they enhance survival and reproductive success. Quine rejected the hypothesis that significant semantical and grammatical aspects of our mastery of language, those which go beyond the mere innate ability to speak, are "hardwired" components of our genetic make-up. Yet, it is one thing to argue that one or the other innateness hypothesis is implausible, or at odds with all the available data, and quite another thing to group the family of "innateness" hypotheses alongside telepathy (Quine, 1969d, p. 306).

[21]In a similar vein Quine wrote: "Granted, the linguist will end up with unequivocal translations of everything; but only by making many arbitrary choices – arbitrary even though unconscious – along the way (Quine, 1969b, p. 81).

flimsy "analogies". Quine held that the interdependent system of analytical hypothesis used in full translation are "unverifiable" whenever they are not directly correlating observation sentences (Quine, 1960b, p. 72). If this is the correct interpretation of Quine's argument in support of the indeterminacy of translation, it would not carry great weight. We need to make a distinction between (a) internal and local analytical hypotheses, like "Gavagai stands for 'rabbit' "; (b) auxiliary hypotheses adducted from other sciences, or previous successful translations; and (c) normative constraints on translation, like the principle of charity, or Lewis' "Manifestation Principle", i.e. the assumption that the native's beliefs normally are manifest in her dispositions to speech behavior (Lewis, 1974, p. 115). There is nothing arbitrary in the selection of auxiliaries in the latter sense. My concern is with (b). While it is true that auxiliary assumptions "imposed" by the translator are not "demanded" by the direct evidence they are nevertheless well-tested and as rationally warranted as any other accepted scientific hypotheses. It is not *rational* for a researcher to disregard evidence gathered from other cultures. Auxiliary hypotheses are not optional in the task of radical translation. The translator has no choice as to whether to adopt the hypothesis or not in the light of the *total* evidence available to her. I expect that Quine would have scoffed at the 'thick' notion of rationality I am relying on in translational matters. Yet, intuitions of what is or is not rationally demanded is hardly a sufficient basis for grounding a thesis that has so many counter-intuitive consequences as the indeterminacy of translation.[22]

There is a second point to be considered. It is surprising that there *is* a translation of the native's sentences, among the (hypothetical)

---

[22] David Lewis, in an examination of Davidson's views on indeterminacy, claimed – in line with a metaphysical realist outlook – that *every* case of indeterminacy of "beliefs, desires, and truth conditions" is a case of insufficient constraints and a challenge to discover additional constraints until the indeterminacy is resolved (Lewis, 1974, p. 118). (His response here is interestingly similar to that of fellow realists like Einstein and Bohm with respect to the postulation of irreducible probabilities in the orthodox interpretation of quantum mechanics.) Lewis' ample use of "principle" should not distract from the fact that the list of constraints and auxiliary hypotheses proposed by him are to a high degree *ad hoc*, and the more are needed, the more arbitrary they appear.

manifold of alternatives, which makes the native speaker's behavior, needs, wants, concerns and beliefs so very similar to our own psychology. The translator cannot expect *a priori* to discover such a manual: the "arbitrary choices" she makes in setting up a translation manual may or may not lead to a consistent and, in point of prediction and ease of communication, successful manual. The fact that "standard" translations seem always to exist is in need of explanation. Quine's view does not easily lend itself to account for this fact. The best explanation for the success of standard translation manual is, analogous to realist's interpretation of a successful hypothesis, that the empirical (sociological) hypotheses the translator "imposes" characterize *correctly* aspects of the native's world picture. The success in translating an extinct ancient writing system shows that independently tested and confirmed hypotheses can effectively narrow down the manifold of translation manuals to one – one which is compatible with the four defining requirements and is *rationally warranted*.

However that may be, it is doubtful that all the constraints mentioned reduce the multiplicity of translational schemes to one. Indeed, Quine suggested that even in the presence of independently justified analytical hypotheses there would be "a lot of slack [...]" (Davidson and Hintikka, 1969, p. 314). The problem then in establishing genuine indeterminacy in translation is to show that the residual "slack" is distinguishable and significantly different from residual vagueness and inductive uncertainty. The following section will examine some examples and arguments to the effect that considerations of auxiliaries in translations are beside the point.

6.3 Empirical Investigations of Cognitive Meaning

In this section I want to examine the stronger claim that even if the auxiliary hypotheses on translation considered in the previous section were taken into account, translation between languages may still be empirically underdetermined. There is an ineliminable freedom (arbitrariness) in attributing theoretical beliefs and a particular reference scheme to a speaker. What a speaker refers to with the expressions of his language is bound up with his theory of the world. The speaker's theory can be variously interpreted or translated while conforming to all dispositions of (verbal) behavior. Hence the necessity to relativize reference to a translation manual.[23] If true, this fact would strongly support the claim to *indeterminacy* of translation. To this end, I begin by sketching Carnap's methodology in Carnap (1955) for empirically investigating what a speaker refers to when he uses a proper name or a predicate.

A *Carnapian inductive linguist* determines by trial and error procedure the extension that a speaker assigns to predicates applicable to observable things or events (to begin with). The linguist directs the speaker's attention to an object selected for typicality, records affirmative and negative answers, and subsequently forms a testable hypothesis beset with the usual inductive uncertainties. *Ostensive reference* is the natural starting point to determine by empirical means what a native speaker refers to when using a mass noun in the translation situation. An indefinite extension of the cross-examination, that is in the limit of a well-designed infinite sequence of varied external sensory stimuli, would lead to the determination of the extension of any observation predicate that the speaker uses. In this way the linguist can make sure that the native indeed refers to the act of "bicycling", say, and not to two rotating wheels, the motion of the legs or any other aspect. Our Carnapian linguist faces *weak underdetermination*

---

[23]There is a disagreement whether such arguments strictly eliminate the notion of reference, or whether the moral is to relativize reference schemes. The first option is suggested as a reading of Quine's thesis in Field (1975) and is endorsed and defended by Stephen Leeds. Leeds' interpretation of the 'inscrutability' of reference thesis appears to be inconsistent with Quine's insistence that ontology is "relative" for speaker's of the *same* language. Quine did nowhere, to my knowledge, claim to have eliminated the notion of reference as a "causal-explanatory" one, as Leeds reads him. Quine claimed that the relation between world (non-linguistic entities) and language – the grounding of language – is not given by the reference of linguistic expressions to external objects etc., but indirectly through the global "amorphous welter" of sensory stimuli. Quine's view of reference contrasts sharply with, for instance, Frege's claim that the sense of an expression determines its reference and Searle's "rules of reference" Searle (1969).

with regard to the extension a speaker associates with an observation predicate in the sense discussed earlier. I have put various oversimplifications aside, like the existence of certain intrinsically vague observational predicates; predicates like "is a *leitmotif*" for which a *definition* is more appropriate; the reference of terms like "voting". And the sequence of the speaker's signs of assent or dissent does not allow the linguist to discern whether the speaker uses the word as delimitating instances in the extension of a one-place predicate (i.e. "this is blue") or whether she uses the same word as an element in the extension of a second-order predicate (i.e. "blue" is a color). I am disregarding complications by the fact that many acts of ostensive reference are not to objects proper but to *states* of objects: eating, drinking, staring, running away, being angry, being tall, being dangerous, being rich, etc. The method is, of course, in line with the classical empiricist theory of associative learning. According to this theory, a speaker learns observation predicates, the primitive predicates of our language, by being repeatedly exposed and directed to exemplars and counter-exemplars by ostensive gestures and then to form spontaneously the appropriate "intensional" hypothesis which is subsequently revised if necessary by trial and error. The inductive linguist's method works backward through the learning processes.

An observation predicate's *intension*, on the other hand, appears strongly underdetermined relative to the set of observations which allowed the determination of the predicate's extension. The meaning or *intension* of a one-place predicate, following C. I. Lewis (1943), is the set of all *conceivable* things which satisfy the predicate.[24] Knowledge of a predicate's intension fixes its extension (trivially) but not vice versa. The empirical determination of the cognitive meaning that a speaker assigns to a predicate seems hopeless. The inspection of all actually existing things will not suffice to discern differences in intension. There is a *natural way*, however, to extent the previous observation basis. By describing various *contrary-to-facts* situations and non-existent objects, perhaps with visual aids if the gradually growing shared lin-

---

[24]Carnap's conception of intension coincides with Lewis' when it comes to the question of testing "intensional hypotheses" (see Carnap, 1955).

guistic resources permit, the Carnapian linguist can elucidate signals of affirmation or dissent from the speaker as answers to her queries. The speaker's reactions to such means as pictures, icons or models of fictive objects (e.g. of a unicorn to probe the meaning of "horse") are admissible linguistic evidence, since it is just a matter of technology to render the "pictures" as life-like and realistic for the speaker as necessary to make them indistinguishable from real objects. Quine objected that it is the *linguist* who estimates what the native would utter under the contra-factual circumstances, yet, he agreed that in field practice the linguist would simply record what the native informant answers when appropriately queried (Quine, 1960b, p. 35). Relative to the *extended* observation basis then intensions of observation predicates are subject to *weak* not strong irresolubility. As with the hypothesis concerning a predicate's extension there is just the residual inductive uncertainty or vagueness of intension (Carnap, 1955, pp. 239f). Carnap thus felt justified in claiming that the attribution of a specific intension for every observation predicate is a testable empirical hypothesis about the speaker. Despite the strong common sense appeal of the foregoing method, it does in effect little to rein in observationally indistinguishable reference schemes. Several reasons have been advanced for why there are schemes which cannot be discerned by the Carnapian inductive linguist on the basis of overt behavior.

Reference by *ostension* is an important means of word-*learning* yet it cannot settle reference uniquely and absolutely, as many thinkers have pointed out. Wittgenstein wrote that the subject to whom an ostensive act is directed to "might equally take the name of a person, of which I give an ostensive definition, as that of a color, or of a race, or even of a point of the compass. That is to say: an ostensive definition can be variously interpreted in *every* case" (Wittgenstein, 1958, rk. 29). Many more such examples are familiar to anybody who traveled a foreign country or learned a foreign language. Quine added that a pointing gesture may be an (admittedly rare) act of indirect or deferred ostension, like pointing to the gas gauge instead of the amount of gasoline in the back of the car. The Carnapian linguist grants all this but believes that her empirical method can gradually reduce such uncertainties about the object of an ostensive gesture by cross-examination (to the point were any hypothetical remaining misidentifications have no practical consequences).[25]

A more telling objection is that reference fixing by ostension only works if both speaker and interviewer share a basic, common *conceptual framework*. Without prior assumptions even signs of assent or dissents are up for grabs, and standards for what is and what is not perceptually salient in a shared situation are lost. There is no self-interpreting behavior. Cross-examination, as a method, does not get off the ground without a minimum of background assumptions about the native. This strongly supports the view that ostension, and hence reference, is an essentially relative affair, and particularly ill-suited to shed light on radical translation. Thus Wittgenstein observed that one already has to "be master of a language" in order to unambiguously be informed by ostensive definitions (ibid., remark 33). The Carnapian linguist, however, could reply that there is nothing unusual in the situation. A physicist confronted with a new phenomenon will first seek an explanation within certain time-tested explanatory models. If the approach is successful, few would object that the explanation is relative to the prior assumptions and model schemes. Analogously, the basic interpretative framework consists of working hypotheses which are frequently independently tested and which may be revised in the light of future findings on the native's verbal behavior. (She can, for instance, draw on psychological research suggesting that the object category is innate, a discovery that would partially fix the framework in which to interpret pointing gestures.) The framework-"relativity" in the case of ostension to gross bodies does not endanger the objectivity of the attribution of a reference scheme any more than it did the objectivity of the physicist's explanation. The analogy is of course contested by Quine. We have apparently come back full circle to the question that had occupied us in the previous section: the epistemological status of auxiliary assumptions in translation. The consideration that moved Quine to claim "ostensive indistinguishabil-

---

[25]Carnap could cite the widespread belief amongst anthropologists that the pointing gesture is the 'truncated' grasping motion of the hand. If this is indeed the origin of ostension, it would rule out Quine's deferred ostension as 'ostension' only by name.

6.3. Empirical Investigations of Cognitive Meaning 235

ity" (Quine, 1969b, p. 39) and that is plausible in its own right, is that reference-by-ostension is necessarily keyed to the perceptual characteristics of an object and blind to its *theoretical* characteristics. The indefinitely continued cross-examination that the Carnapian linguists imagined cannot resolve this kind of empirical irresolubility: no test can discriminate between empirically equivalent theories. Sentences describing theoretical features of spatial compacts are subject to the underdetermination, if not indeterminacy of translation. "I use stimulations in meanings [...]. Reference, even on the part of observation terms, is in my view theory-enveloped and thus subject to the indeterminacy of translation." Davidson too sees in the indeterminacy of translation a ground for the inscrutability of reference. Putnam thinks the *reverse* holds (Putnam, 1975b, p. 179), based on inferences from suggestive examples, for instance, the existence of isomorphisms between domains of objects. G. Massey constructed empirically irresoluble translation manuals based on the duality relation in first-order logic (Massey, 1978, p. 51).[26] I think Quine was moved by both kinds of arguments (compare Quine, 1969b, p. 35; 1992b, p. 8). The general, methodological critique of the Carnapian linguist, however, appears to end unsatisfactory in a circle. This leaves us with examples of sentences and beliefs which undermine reference fixing by ostensive acts even in the simplest cases.

If we follow Quine, the Carnapian linguist wrongly believes words are directly keyed to the stimuli which trigger a speaker's verbal dispositions, not sentences "holophrastically". The basic linguistic entity is, however, the *sentence* and in particular the observation sentence. The disposition to affirm an observation sentence is correlated as a whole ("holophrastic") to the total neural sensory input, a whole that is only syntactically structured. The external "circumstance" of the observation is summed up in the total stimulus input on the speaker's receptors. The total stimulus and not the object directly triggers the speaker's disposition to affirm. (The impact of the stimulus is *publicly*

---

[26] Another, unusual example is the Ramsification of abstract objects, discussed above, which "brings out indeterminacy of reference [...] by waving the choice of interpretation" (Quine, 1995, p. 74).

observable to a degree: one can observe that the speaker watches something, the direction of her gaze, the duration of the act of looking, etc.) In the radical translation situation, consider two interpretations of predicates and objects *below the sentence level* that respects syntax and leaves the speaker's disposition to affirm the observation sentence whenever appropriate intact. Permutations within domains, "proxy-functions" and compensatory reinterpretations on the sub-sentence level preserve the truth (affirmation) of sentences, see chapter 4. The semantics below the sentence level is *underdetermined* by the speaker's sensory input. The Carnapian linguist, who registers the circumstances, the utterance and the affirmation, cannot infer from this information alone the speaker's semantics on the sub-sentence level.[27]

> We begin to appreciate that a grand and ingenious permutation of these denotations, along with compensatory adjustments in the interpretations of the auxiliary particles, might still accommodate all existing speech dispositions. (Quine, 1969b, p. 48, II)

What a speaker refers to in making a sincere assertion, hence what the speaker takes to exist, is not an absolute, language-transcendent objective matter. Empirically meaningful talk about reference presupposes a largely unrestricted *decision* how to render the speaker's sentences in one's own language. Between two speakers sharing a language and relative to the "homophonic" translation manual (the identity map) reference is explicated deflationary in the style of Tarski's convention T.[28]

By way of closing this section let me review the situation we have arrived at. We have started out in the previous section by examining the underdetermination (by overt behavior) of systematic attributions of beliefs, cognitive sentence meaning, propositions, etc. to a speaker

---

[27]Curiously, Carnap had noted the difficulties created by the "permutability" of the basis in *Der Logische Aufbau der Welt* (1928, § 154).

[28]The home language's reference scheme serves, according to a plausible interpretation defended by Leeds, as an (arbitrary) fixpoint relative to which referents can be assigned to the *native*'s sentences, much like a location on a map is determined relative to a coordinate system.

in the radical translation setting. The claim that this case of observational irresolubility signals an indeterminacy, an absence of fact in what a speaker believes "absolutely" speaking, turned on the epistemological status of auxiliary hypotheses about cognitive processes drawn from linguistics, neurophysiology, or anthropology. It turned out that Quine would grant such "analytical" hypotheses a status closely similar to a Reichenbachian convention: effective in bringing about a unique translation, but without genuine factual or explanatory content. The present section focused on the possibility to determine by observation the denotation and intension of a native speaker's observational vocabulary, i.e. an important subset of her vocabulary. The Carnapian linguist shares with Quine the belief that language is an essentially social construction. We can thus hope to side step the issue of the epistemological status of linguistic or neurophysiological auxiliaries. It turned out general, methodological objections against this method were circular. Reference fixing by ostension functions only by making substantial prior assumptions about the speaker (though not of the neurophysiological kind), since reference is "theory-enveloped". This brings us back to Quine's critique of the status of auxiliaries. Suggestive examples of variant reference schemes compatible with all observable evidence, the second pillar for the claim of indeterminacy, have all a formal 'symmetry' structure resembling isomorphisms. It is natural to interpret these examples as exhibiting an unsurprising degree of conventionality in denotation (no attempt at determination can do better than up to isomorphism). In sum, indeterminacy does *not follow* straight from observational indistinguishability. The inference to an "absence of fact" in an empirically irresoluble dispute depends on additional substantive input; in the case of semantic irresolubility the input concerns the nature of mental entities in question. This is not the place to discuss the merits of these assumptions in full, the point is that the assumptions are crucial. If I am right on this, this dependence changes the philosophical significance of indeterminacy.

6.4 Indeterminacy and the Absence of Fact

My take on the phenomenon of indeterminacy in our language differs from that developed by Hartry Field. He distinguishes various types of examples of indeterminacy (Field, 1974; 2000):

- Ordinary vagueness (it is indeterminate whether Otto is poor, or whether a given molecule falls within the boundary of Otto's body)

- Indeterminacy of reference of physical predicates in superseded theories with respect to their counterparts in successor theories

- Indeterminacy generated by certain structure preserving mappings

- Quine's "gavagai-rabbit" translational indeterminacy

- Indeterminacy in the notions of "set" and "$\in$" relative to the axioms of ZF

One may add to this list indeterminacies in the attribution of desires and propositional attitudes.

Field developed two approaches for analyzing indeterminacy. First, in the hope to show that semantic theory is a sound, autonomous scientific enterprise despite Quineian indeterminacy arguments, he put forward a semantics that would explain indeterminacy in some of the cases listed above. This generalization of the usual notion of denotation is inspired by the intricacies of accounting for vague predicates. According to Field's semantics the terms of a language have or are capable of having "partial denotation", i.e. predicates can *partially denote* in nonoverlapping sets of objects, and quantity terms can denote more than one physical quantity. The *Newtonian* term 'mass', for instance, can be identified with either rest mass or relativistic mass in relativistic particle dynamics. Instead of identification Field speaks of 'translation' of the Newtonian term into the relativity theory. Choosing one of

6.4. Indeterminacy and the Absence of Fact

the alternative translations makes certain hypotheses of Newton true while others come out false, and *vice versa* for the alternative translation manual. Choice between the alternative translations is objectively indeterminate, according to Field (1974, p. 206). He proposes that Newtonian 'mass' partially refers to relativistic mass and partially refers to rest mass.[29] (The partial denotations, at least in simple cases, are taken to be fixed by causal processes.) Clearly, if partial denotation *is* the basic way terms refer than we should never have been surprised by indeterminacy, never have treated cases of indeterminacy as exotic exceptions. The theory explains the phenomenon indeterminacy, and how physical concepts refer trans-theoretically, across epoch-making scientific upheavals. I cannot do justice here to Field's intricate theory. But note that the partial denotations of a predicate or quantity term like 'mass' are strongly context-sensitive. 'Rabbit' will not only partially denote the set of rabbits and the set of undetached rabbit parts, but in addition all other possible assignments compatible with admissible translation manuals as well. This includes the 'cosmic complement' of the very rabbit. Similarly for 'mass' and its role in either observationally equivalent alternatives to *current* relativity theory, or in future theories of kinematics insofar as they still have a term 'mass' among its primitives. I find it difficult to decide whether semantics centered on a contextual notion of "fuzzy" denotation like Field's is much to be preferred over Quine's (and Leed's) outright abandon of the idea of an "autonomous" scientific semantic theory. Does this theory really offer an explanation of the linguistic phenomenon of indeterminacy, or is it just a paraphrase in a new vocabulary? Quine used underdetermination arguments to cast doubt on the possibility of semantic theory, in the same way R. M. Hare had argued against a theory of objective moral values. My objective is to examine in how far the inference from underdetermination to indeterminacy is justified, i.e. whether positing indeterminacy of terms in a language is a method to avoid potential skeptical or anti-realist consequences. Field *grants* all kinds of cases

---

[29]The application of Field's semantics to the radical translation of "Gavagai" is straightforward: "'rabbit' partially signifies the set of rabbits and partially signifies the set of undetached rabbit parts" (Field, 1974, p. 216).

of indeterminacy, see the list above, and hence essentially agrees with Quine on this point. He differs on the consequences for the status of semantic theory, the prospects for establishing a scientific semantic theory. The question I raise is prior to that, i.e. it regards the justification for positing indeterminacy in cases of empirical or rational irresolubility. (This is an epistemological, not a semantic problem.)

Field's second contribution to our topic is a proposal to represent a rational agent's beliefs in conflicts of assertions where she suspects that there is no fact to the matter by non-standard degrees of belief. First, consider the case of two incompatible hypotheses H_1, H_2 which agree on past evidence and which we find difficult to decide empirically. Our beliefs, measured by Bayesian probabilities, satisfy the relation: $prob(H_1 \vee H_2) = prob(H_1) + prob(H_2) - prob(H_1 \& H_2)$. Assume for simplicity's sake that there are no further alternatives to take into account; then one has $prob(H_1 \vee H_2) = 1$. It is certain that one of the two is true, the other correspondingly false, and the respective probabilities lie anywhere between 0 and 1, depending on past performance and prior estimates. (If such estimates are lacking, the agent's subjective ignorance would best be represented by assigning the value 0.5 to each alternative.) Yet, if the two are empirically irresoluble and, furthermore, if one has reason to think that it is indeterminate which of the two is true and which is false, an ideal rational agent cannot well assign different probabilities to the hypotheses. Nor can she assign numerically equal probabilities: as a way to represent the vagueness of "Otto is poor" and "Otto is not poor" (if the issue arises, that is) the suggestion seems clearly wrong. A good analogy to Field's strong interpretation of vagueness, on which this reasoning rests, is the objective indeterminacy of complementary states in quantum physics (Heisenberg's "uncertainty relations"). Assigning degree of belief 0.5 to each alternative would put vagueness on level with perfectly determinate predicates. The equal-probability proposal is a poor description of vagueness (in the strong interpretation). Although an agent's assignments of classical (i.e. Bayesian) degrees of belief to each individual hypothesis are problematic when she suspects choice to be indeterminate, at the minimum she would have to maintain that $prob(H_1 \vee H_2) = 1$ for the disjunction and $prob(H_1 \& H_2) = 0$ for the

6.4. Indeterminacy and the Absence of Fact

conjunction of the two rival hypotheses.

Besides vague predicates the following illustration borrowed from Robert Brandom is typical for cases Field thinks are exemplary of indeterminacy: suppose a community of mathematicians uses the signs / and \ in contexts where we would use the imaginary unit i or $-i$. There is nothing in the mathematicians' practice to indicate whether the translation manual (A) $/ = i$ and $\backslash = -i$ into our symbolic system, or the manual (B) $/ = -i$ and $\backslash = i$ is the objectively correct one. The statement $/ \cdot \backslash = 1$, like any other complex equation, comes out true under either translation. Field holds that Brandom's example of rational irresolubility is an instance of indeterminacy of translation, and proposes to represent our "wavering" epistemic attitude towards the question whether manual (A) or (B) is correct in the following way:

> The indeterminacy consists in the fact that if we do that [introduce new names for the square root of -1 – T.B.], our degrees of belief in '$/ = i$' and '$\backslash = i$' should each be 0, even though the degree of belief in the disjunction should be one. (Field, 2000, p. 303)

Indeterminacy in the rational choice between alternatives is, according to Field, best expressed by the agent's assigning the degree of belief 0 to each alternative and the value 1 to the exclusive disjunction of the alternatives. The agent's degree of belief in either one need not sum up to 1 when choice is arbitrary; in fact, the difference to the limit value 1 measures the agent's belief that the choice is indeterminate. In my schematic example above, none of the two hypotheses is to be preferred over the rival one. Instead of the "ignorance value" 0.5 genuine indeterminacy requires the rational agent to assign a degree of belief close or equal to 0 to H_1 and H_2. The agent rejects H_1 *without accepting* H_2. This reflects the agent's belief that there is no fact to the matter which is true and which false. Does a coherent belief function Q, say, exist that satisfies these exceptional requirements? Field shows that for a quantifier-free language L, equipped with a necessity-like operator "determinate" (D) satisfying the axioms of the modal system $S4$, and additional axioms relating *prob* and Q, the answer is *affirmative*.³⁰

Understanding indeterminacy may require an even more radical break with classical thought on rationality than abandonment of classical degrees of belief. Returning to my initial example, since H_1 and H_2 are exclusive rival accounts of the relevant data and syntactically incompatible one hypothesis is objectively true and the other false (although we do not know which is which). Yet, if there is reason to believe that the empirical irresolubility of the two is a case of genuine indeterminacy one wants to deny precisely this: that one alternative is objectively true as a matter of (unknown) fact and the other correspondingly false. Thus a genuine indeterminate choice between alternatives is a context where the *law of the exclude middle* is on the line. Field, like Putnam, motivates the departure from classical logic (bivalence) by an analysis of the vague predicate "rich" (Field, 2000, p. 292). Rejecting the law of the exclude middle can mean two things: (a) either the adoption of a strongly non-classical logic for modeling reasoning with indeterminate statements (vague predicates), or (b) refraining from assertion or rejection of the disjunction $H_1 \vee H_2$ in contexts where indeterminacy looms.

The suggestion to "give up" classical logic to solve a conceptual problem comes up with great regularity: Reichenbach and Putnam advocated it as a way to "improve" the standard interpretation of quantum mechanics; Putnam and Field suggest alternative logics as the proper treatment of languages with vague predicates; and Putnam and Field contemplate rejecting bivalence for the case of indeterminacy (to mention just a few thinkers).³¹ I think the temptation should be

---

³⁰Among the atomic sentences of L are those like "Otto is poor" which apply vague predicates to objects. For a language where all atomic sentences are determinate, the function Q coincides with Bayesian probability assignments. The notion of indeterminacy appears to require that no new evidence E can drive the initial value of $Q(H_1)$, say, up or down. The evidence need neither be "empirical" nor be in the deductive content of the hypotheses under consideration. Formally, the "indeterminacy" condition on Q requires strong independence from any evidence: $Q(H_n|E) = Q(H_n)$ for every n in the exhaustive range of mutually incompatible, genuine alternatives and for every (observation) statement E.

³¹Putnam advocates treating "vague predicates [...] just as undecidable predicates are treated in intuitionist logic" in Putnam (1983c). "Fuzzy logic" systems,

resisted. The rejection of the *tertium non datur* is *ad hoc*; nothing in the sciences requires it, not even quantum mechanics. Moreover, it is un-illuminating as a means of formulating indeterminacy. The move blurs the crucial question: under what conditions is one justified in positing indeterminacy as the ground of empirical (rational) irresolubility? If Field has wider, substantive reasons for adopting one of the non-classical logics into our language he does not make them explicit. One thinker who does offer such reasons is Dummett and they have strong verificationist underpinnings (he says little about the specific issue of vagueness to my knowlegde). My point is simply that the interpretation of observational (or rational) indistinguishability in terms of a non-classical logic requires substantial semantical or metaphysical premises in its defense. It is neither "natural" nor inevitable. The merits of Field's interpretation (and he does not fully embrace it) are to be judged entirely on the merits of these premises.

I return to Field's proposal to adopt a non-standard degree of belief function instead of a non-standard logic. My disagreement with this approach to understanding indeterminacy starts with the interpretation of Q as a *belief* function. The requirement $Q(H_1) = 0$ reflects the agent's belief that H_1 is *false* as an account of the facts. This is at odds with what Field wants to say, namely that there is nothing for the agent to be right about in case of indeterminacy. He holds that the straightforward interpretation of $Q(H_1) = 0$ is mitigated by the agent simultaneously rejecting H_2, the only alternative (the negation): no hypothesis is privileged. I do not see how the simultaneous low degree in the alternative changes the basic interpretation of Q applied to H_1, say. Field has shown that one can *relax* the requirements for a degree of belief function *prob* so as to admit that the agent can assign the value 0 to exclusive alternatives. It is of course *prima facie* surprising that the agent can have such a set of beliefs, but the very coherence of Q and its coincidence with standard probabilities in the sub-language of determinate sentences shows that Q is a belief function. Indeterminacy then is not satisfactorily represented by the simultaneous requirements $Q(H_1) = Q(H_2) = 0$ and $Q(H_1 \vee H_2) = 1$.

---

too, have been candidates for modeling vagueness.

If I am right in this, then the prospects for modeling indeterminacy in terms of degrees of belief are dim. A further consideration supports this view. Indeterminacy is partially characterized by a range of (non-isomorphic) mutually incompatible alternatives all empirically indistinguishable; it is never present when we are dealing with an (essentially) finite range of alternatives. More precisely, indeterminacy is not present if finite alternatives are empirically irresoluble. "Finite" irresolubility is more plausibly understood as a case of ignorance than objective indeterminacy. This is a consequence of the universality thesis of empirical indistinguishability (U), discussed earlier see section 1.2. I will defend my privileging finite alternatives, which is controversial in view of Field's examples, in a moment. To go on then, standard theories of degree of belief require in cases characterized by countable many alternatives that the numerical assignments to individual hypotheses are not all equal (since their infinite sum does not converge). The agent thus is forced to play favorites in circumstances where she believes no one choice is more rational than the other. Indeterminacy is just such a context. While in determinate contexts there is a rationale for accepting "playing favorites" (the priors are irrelevant in the long run, etc.), when facing indeterminate choices the agent cannot not be right in his subjective assignments, since there is nothing to be right or wrong about (and the long run does not matter). The moral I draw is that standard theories of rational choice simply do not apply to the indeterminacy of translation, etc. They are not adequate for modeling an agent in the face of genuine indeterminacy.

The examples of indeterminacy Field draws upon in formulating his proposal are binary alternatives, or instances of vague terms in our language. I take them one after the other. First, the dual translation manuals of the imaginary unit, as in Brandom's example, are isomorphic and potential notational variants. Horwich's analysis, explained above, fully applies. It is more plausible that the imaginary unit is defined by its place in a certain mathematical structure, i.e. that reference is determined structurally, than to assume that we are faced with inscrutable reference to an abstract object. Yet, granted that these translation manuals are genuine alternatives (not mere notional variants) I think we should bite the bullet and admit to ignorance

6.4. Indeterminacy and the Absence of Fact

rather than to indeterminacy. The inference to an indeterminacy requires more, for instance, a nominalist or constructivist view of the nature of mathematics. Second, examples and considerations of vague predicates figure large in Field's thinking on the matter of indeterminateness. There is nothing in our practices that fixes the reference of 'bald' for borderline cases like Otto's scalp. The extension is indeterminate. On the contrary, I see few informative connections between the indeterminacy issue and the nature of vague predicates. (1) Even if all predicates in our language were determinate the issues of indeterminacy of radical translation, the underdetermination of the sciences, etc., would persist. So vagueness cannot be of the essence in assessing the indeterminacy. (2) Nor does vagueness make a good model of indeterminateness. Vagueness, opposed to empirical irresolubility and indeterminacy is eliminable and negligible. It is eliminable, since although one may not be sure whether Otto is poor, one can rank Otto's poorness relative to that of Gustav's and Alex's. This narrows down the extension until practical standards are met, see Quine (1960b, p. 127) who suggested that vagueness does not really matter ontologically.[32] Vagueness is a phenomenon of ordinary language that is absent in the mature exact sciences. If the truth of any hypothesis depended on Otto being bald or not an exact measure would have been found. But since little by way of objective knowledge depends on whether Otto is bald, or whether a certain molecule belongs to his body, local contextual conventions will settle the matter.[33] In sum, the issues of empirical irresolubility owe nothing to the phenomenon of vagueness in ordinary language (I leave it open whether it really poses a difficulty for metaphysicl realism, as Putnam claimed).

---

[32] Putnam though, taking up a suggestion by Davidson, argues that vagueness is one more problem for a "metaphysical realist" point of view, since the truth-predicate itself is vague Putnam (1983c).

[33] On this matter I find myself in agreement with Quine: "What I call my desk could be equate indifferently with countless almost coextensive aggregates of molecules, but I refer to it as a unique one of them, and I do not and cannot care which. Our standard logic takes this also in stride, imposing a tacit fiction of unique though unspecifiable reference" (Quine, 1995, p. 57).

I return to the problem of when to posit indeterminacy in cases of empirical irresolubility. If one can justify that a conflict of assertions is indeterminate, the conflict is resolved: the dispute is shown to be 'empty', there is no fact to which alternative is true and which is false. Positing an 'absence of fact' thus aims – by less problematic means – at delivering the same deflationary result in the face of empirical indistinguishability that criteria for lack of cognitive meaning and the existence of pseudoproblems sought to deliver a century ago (compare section 5.4). I will now examine two proposed criteria for when the posit of indeterminacy is justified. Recall that the concept of empirical irresolubility can be tightened along many dimensions. The *first proposal*, naturally, is to take observational irresolubility itself as a sufficient and necessary criterion of the absence of fact. There is no restriction imposed as to the when or where of relevant observations in space and time, but the inferences based on these observations (or Quine's observation "categoricals") are strictly hypothetico-deductive in classical first-order logic, not probabilistic.[34] Observational indistinguishability means that there is no crucial observation, in analogy to crucial experiments, that would indicate which of the alternatives is true, which is false. In typical cases one can prove the equivalence of empirical deductive content and hence knows *a priori* that a crucial observation is not possible. (A stronger version – required if no such proof is available – demands that the range of alternatives be observationally irresoluble in every close possible world, i.e. worlds compatible with the truths of physics in the actual world.) Moreover, the observation is to be unaided, without amplifying instruments. This latter requirement is, of course, highly unrealistic when the alternatives are proper scientific theories. My concern here is to state a basic variant of the criterion. The proposed criterion fits the standard examples of *semantical indeterminacy* by Quine and Field. It has a distinctly epistemological flavor: it dictates that if the alternatives – which we assume to disagree about a common posit, (or term) its existence and

---

[34]Reichenbach, who advanced probabilistic thinking in semantics and induction, limited its scope to empirically decidable questions. This is for the most part a consequence of the frequency interpretation he had adopted.

6.4. Indeterminacy and the Absence of Fact

properties – do not differ on what one can *observe* then their theoretical differences do not possibly reflect what is and what is not. Even steep differences in comparative simplicity cannot objectively single out one as true. If there cannot be a crucial observation whose outcome would determine which alternative is true and which is false – quite apart from the usual Duhemian uncertainties – then there is no fact to the matter:

(I) There is no fact to the matter which in a spectrum of mutually exclusive, exhaustive, genuine alternatives is true if and only if the alternatives are irresoluble with respect to the totality of *observation* sentences or observation categoricals.

(I) presupposes that singular observation sentences are relevant for judgments of truth and falsity, and that the observation basis is the *maximal*, ultimate data basis on which to make this judgment. Without the first assumption the criterion has unwanted consequences, for example that the debate about the continuum hypothesis is "empty" (the debate whether CH or $\neg CH$ *may be* empty – but surely not for the reason that observations do not make a difference). The maximality assumption requires that the observation basis cannot be "trumped", that there are no additional facts, generalizations or laws (inside or outside the domain of application of the conflicting hypotheses) that could resolve the empirical irresolubility. Without the second assumption observational irresolubility is subjective and trivial in point of epistemological inquiry. By choosing a basis that is suitably impoverished any set of alternatives becomes "irresoluble". The relevance and maximality premise are rarely stated explicitly in the literature on indeterminacy and underdetermination, but they are crucial. Quine held as we have seen, for instance, that observation of overt behavior is essential to the interpretation of utterances of the native speaker. Everything else the translator wants to bring into determining meaning and reference of an utterance is an "analytical hypothesis" – not a fact. Proposal (I) for a sufficient criterion hinges on the strength of the maximality premise. In unpacking the criterion I suppress the dependence on the hypothetico-deductive methodology:

HD 1 Suppose a spectrum of mutually exclusive, exhaustive, genuine alternatives is observationally irresoluble with respect to the totality of observation sentences or observation categoricals.[35]

HD 2 The totality of singular observation sentences, or observation categoricals, is potentially relevant for the truth or falsity of the alternatives.

HD 3 No other scientifically acceptable truths beside true singular observation sentences, or observation categoricals, can determine truth or falsity of the alternatives.

HD 4 From 1., 2., and 3. conclude that there is 'no fact to the matter' which of the alternatives is true and which is false (the conflict is "verbal").

The strength of the last conclusion depends on how strongly the conflicting assertions are (metaphysically, epistemologically) tied to observation sentences, i.e. on premise 3. The sharper the formulation of the third premise the clearer it becomes that the premise is a *petitio principii*, reducing the argument to a tautology. The moral is that observational irresolubility by itself is not a sufficient reason for concluding that the choice is indeterminate. (Should one, instead, say that indeterminacy is the *best explanation* for an epistemological quandary characterized by 1. and 2., if not 3.? Field, for instance, uses the phrase of "positing" indeterminacy. Posits and postulates are catchwords borrowed from a picture of the sciences that describes the scientists' activities as problem solving by postulating entities along with laws and drawing out hopefully useful consequences in a controlled way. "Positing" an *absence of fact* to account for the *failure* of procedures of justification to determine the truth value is a different matter.)

---

[35] There is, as noted before, some leeway in defining what are "genuine" alternatives. I too tend to dismiss alternatives as genuine, which are "potential notational variants", i.e. that can be formally transformed into each other term by term in systematic fashion.

6.4. Indeterminacy and the Absence of Fact

As noted before, (I) is not sufficiently general in that it does not fit cases where the alternatives are proper theoretical, explanatory accounts of a domain of observation. Those observationally irresoluble alternatives call for probing with measurement instruments in experimental settings. At first sight, this inclusion of amplifying instruments into the revised statement of (I), i.e. (IP) below, may mean nothing more than taking aboard an additional layer of theory, i.e. auxiliary hypotheses describing the instruments, experimental set-ups, standards of statistical error, etc. Quine and others held that this is not a principled difference. Yet, in scientific practice the distinction between principles and auxiliaries is deeply ingrained and mirrors different levels of dispensiblity. This observation justifies treating auxiliaries not just as "more theory". A serious difficulty for formulating a criterion that takes account of the availability of instruments is the notion of "instrument" and measurement itself. One would have to exclude the possibility of a hypothetical measurement instrument that would be sensitive to the controversial "theoretical" claims between irresoluble alternatives. The way to do so is either to demand compatibility with an ideally known, complete corpus of physical laws, or to invoke *a priori* considerations on what measurements are possible. The latter option strikes me as hopeless, while the former makes the explicit inclusion of instrumentally aided observation superfluous by appealing directly to physical truths. It is best to seek an alternative to (I) based on an ideal system of physical laws and facts, as does (IP) below.

The next version of our criterion strengthens the previous one with respect to the fundamental truths that can count for or against the truth of a theory: crucial observations are replaced by determination by the truths of physics. One explication of "determination by the truths of physics" lists the following two requirements: (i) a specification of the regions of space-time of which the primitive predicate of physics are true of, and (ii) co-extensiveness of every non-physical predicate to be determined with a predicate of physics (Friedman, 1975, pp. 357–359). The second clause requires term-by-term reduction of the observationally underdetermined alternatives to the primitives of (today's or an ideal) physics as the science. (ii) is unnecessarily restrictive: neither chemistry (viz. "electro-negativity") nor classical thermodynamics (viz. "entropy") is "determined" by physics in this sense. The reducibility requirement (ii) has no support in Quine's comments on indeterminacy. His notion of determination means more or less implication, logical compatibility or incompatibility. Alternatively then, assume that we have, following Ramsey and David Lewis, an ideal, unique systematization of (a) all singular facts (in the sense of (i)) and physical *laws*; and (b) of all chemical, biological, anthropological, etc., facts and laws (provided there are any). We need not assume that (b)-type facts and laws are ontologically and term-by-term reducible to (a)-type laws and facts. The physicalistic variant, however, either does make this claim about the priority and reductive explanatory power of the (a)-type systematization, or at least claims a supervenience relation between the two realms of law. With the other assumptions (possible worlds, deductivism) as in the previous paragraphs the following criterion suggests itself:

(IP) The dispute over which in a spectrum of mutually exclusive, exhaustive, genuine alternatives is true is empty if and only if none of the alternatives is consistent with the ideal systematization of facts and laws (a) and (b).

As emphasized before Quine, when pressed about his behavioristic methodology and outlook, insisted that the notion of sentence meaning, and propositional attitudes, will be indeterminate under (IP) as well as (I); compare claim (PH) section 6.2. Some commentators formulate the indeterminacy claim with respect to *dispositions to behavior*, namely dispositions to accept or reject sentences (Gibson, 1982, p. 69). The disposition itself is not an observable, therefore it is more appropriate to see this (third) formulation as a restricted variant of (IP), restricted in addition to semantic features of discourse. That a systematization of singular facts and laws (a) and (b) is the the most complete knowledge of fact one can hope for is plausible, and so the *maximality* condition implied in (IP) is satisfied. There is no better standard and ultimate data basis for judging the truth or falsehood of alternatives that aspire to describe objective features of the material universe.[36]

---

[36] If (b) is dropped from the criterion, the last claim becomes more problematic since the alternatives may make statements in terms which may not be definable

The criterion is not readily applicable to "undecidable" mathematical statements – on a view that interprets mathematical truths either as analytic, or in a realist spirit as necessary and *a priori*. The *relevancy* condition is not satisfied, the contingent facts and laws of our universe do not give us the truth value of mathematical statements. And potential heuristic considerations or judgments of fertility in generating proofs or new questions, etc., in relying on CH are not included in the test and data basis of (IP). The relevancy condition *is* satisfied, and hence (IP) applies, if one interprets mathematics (logic, set-theory) as an integral, quasi-postulatory part that enhances the empirical efficacy of the scientific system. In that case, choice between continuum hypothesis and its denial is *not indeterminate*. Adopting the continuum hypothesis or replacing or dropping it "saves the phenomena" to the extent that they are know today equally well. Yet, the dispute is not "empty", if we take the argument from potential use in future theories seriously. Future theories (prompted by new evidence) may in some way call for CH.[37] At any rate, indeterminacy (as noted in connection with (I)) follows only on an extra premise in addition to relevancy and maximality: the truth of an indispensability account of the nature of mathematics. This premise is not uncontroversial.

I have said above that the maximality condition is plausibly satisfied when the basis is an ideal sum-total of physical truths. However, the plausibility felt is strongly informed by our picture of the sciences, a perception of our place in the natural world, and paradigmatic applications of (IP) to controversies between rival theories. For those who hold a dualistic ("substance" or "property") view of the human person, say, or hold that intentionality is an irreducible feature of our representation of the world, the maximality premise is false. On this view, to claim that any dispute is either decidable on this basis or is

---

in, or otherwise be reducible to physicalistic terms.

[37]Quine argued this point in Quine (1973a, p. 65). Field argues that the question whether CH or $\neg CH$ is indeterminate because set-theoretic terms, in particular the notion of "set", are referentially indeterminate (underdetermined by the axioms of ZF) (Field, 2000, p. 304). He does not address the argument from potential use in future theories. As a binary alternative the issue does not qualify as being 'indeterminate' on my account.

factually "empty" skirts the tautological. The relevance assumption too becomes dubious. As a self-standing, sufficient criterion for indeterminacy (IP) fails: its force depends on the strengths of the reasons one can advance for the maximality and relevance premises. The criterion has to be taken as a multiple-premise argument like in the case of (I). Quine and others are committing a *petitio principii* in advancing criteria like (IP) as part of thought experiments in order to refute "mentalistic", linguistic or semantic views of mind and language.

6.5 Quine's Pragmatic Interpretation of Underdetermination

There is an important case of empirical irresolubility that cannot be accounted for by a basic indeterminacy in a term. In this section I examine the significance of the

> omnipresent under-determination of natural knowledge generally. (Quine, 1969d, p. 304)

If the under-determination of natural knowledge, i.e. the sum-total of common sense truths and scientific truths where they do not conflict, by all possible observation categoricals is indeed "omnipresent" than radical skepticism appears to be a natural epistemological consequence. Neither Quine, nor any other commentator I know of, draws this consequence – for very different reasons though.[38] I have reviewed the *methodological* reasons Quine advanced for the universality thesis of empirical irresoluble conflicts on the level of global theory: (a) a literalistic reading of Duhem's thesis, (b) subsequent holism, (c) the bare hypothetico-deductive methodology as the minimal core of scientific methodology from an epistemological point of view, and (d) the

---

[38]Moulines (1997) argues that even in the long run science is likely to be stuck with more than one fundamental theory which are mutually incompatible yet not necessarily empirically equivalent. He emphasizes the difficulties for achieving a full ontological reduction among theories and argues that Peirceian "convergentism" is an utopic goal.

6.5. Quine's Pragmatic Interpretation of Underdetermination 253

alleged function of theory as an instrument of "mass coverage". From these premises it appears to follow that there is an inferential "gap" between what our rich theory of the world postulates and the limited, truncated observation basis that serves to check and justify the theory.

One thinker who denies this and locates the flaw in Quine's notion of observation sentence is Barry Stroud (on Quine's novel notion of observation categorical compare section 3.2). Stroud argued that there are *two* readings of Quine's thesis of *sensory* irresolubility and both are unsatisfactory. In the *first* reading the thesis would assert the existence of a "gap" in the causal sequences of, say, light rays being scattered on objects, hitting the retina, traveling up the optical nerve to regions of the cortex that physically realize an individual's "being disposed to assert" the sentence p. Yet, there can be no gap in a causal sequence of physical events, from a naturalist perspective. In the *second* reading the thesis would claim the *inferential under-determination* of the truths of ordinary discourse by sensory information or evidence, i.e. true observation sentences. In this reading, Stroud claims, Quine is committed (a) to the idea of 'awareness' or 'aboutness' and (b) to the epistemological *priority* of observation sentences. Now, (a) contradicts the claim that our sensory input is 'meager', since an observation sentence like "there are five chairs now in this room" carries a great number of presuppositions and (counter-factual) consequences (unlike a Neurathian report). Moreover 'awareness' is an intensional concept hardly compatible with Quine's version of naturalism. On the other hand (b) introduces the traditional foundationalist conception which leads straight into to scepticism whether that conception is embedded in the naturalistic project or not (Stroud, 1984, p. 248). The effect of Stroud's criticism is to diminish, if not to eradicate the "gap" between rich beliefs about the world and the relatively poor observational basis for these beliefs.

Stroud's *first* objection aims at the ontological and nomological commitments imported in observation sentences about ordinary states of affairs. Since the observation sentence is rich in content, "neural intake" and "torrential output" are not separated by an epistemological hiatus. If there is a residual gap it bears analogy to the gap between a richer and a poorer description. However, Quine's notion of stimulus observability is not tied to any such "absolute" contentual commitments as Stroud notices. An observation sentence is characterized not by what it refers to, but how it is acquired, how widely it is shared in the speech community, and how closely it is tied to sensory stimulations. (The thesis that a speaker's reference is 'inscrutable' for the interpreter illustrates this point.) This analysis of the nature of an observation sentence may be wrong, but Stroud does not give any arguments why we should think so. Stroud's *second* objection is that with the notion of stimulus-observability Quine unwittingly smuggled a foundationalist conception into the naturalistic project. Language and theory are indeed primarily connected to the world by way of external stimuli impinging on the speaker (not by way of a special causal mechanism of "reference") and hence observation sentences have a special status. The elite character of observation sentences does not translate into epistemological priority: their status derives from the sciences (anatomy, physiology, neuroscience, optics, acoustics, etc.). The role of observation sentences is as conjectural as anything else, these sentences lack the classic characteristics of incorrigibility, privacy or certainty. In the methodological circle which is so characteristic of epistemological naturalism stimulus observability is just one way of placing the interface between speaker and world. Davidson, for instance, positions the interface (the "cut") at the surface of those bodies in the environment of the speaker which are direct external causes of whatever stimuli the speaker eventually receives. It does not matter much for the naturalist project where the initial "input" into the speaker is located. I conclude that Stroud's objections against Quine's notion of observability do not stick. (Whether this in turn entails that epistemological naturalism is a satisfactory, non question-begging answer to skepticism is another question.)

Provided then that there *are* conflicting hd-equivalent alternatives to any global system of belief – a point that was argued for by examples, by logical argument ("permutations") and critique of scientific method (Quine, 1975, p. 326) – the question arises how empirical irresolubility is possible at all. Quine maintained that underdetermination is rooted in the way we *theorize* about the world.

> Here, evidently, is the nature of under-determination. [...] Any finite formulation that will imply them [the 'infinite lot' of empirical truths – T.B.] is going to have to imply also some trumped-up matter, or stuffing, whose only service is to round out the formulation. There is some freedom of choice of stuffing, and such is the under-determination. (Quine, 1975, p. 324)

It is the way the material world is that makes it necessary to use theoretical concepts and natural laws in deductively formalized theories for the purpose of scientific systematization and making predictions. The world *could* be less complex than it is so that in order to make predictions it is sufficient to check a finite conjunction of observation sentences, or to run a recursive function in a suitable observation language.[39] However, our world is fairly complex. A theory's potential applications are unlimited and experience is "open" (Heinrich Hertz), hence the utility of an axiomatized theory or a finite system of laws and hypotheses in order to predict and control of the environment. Quine suggested that theoretical concepts are there to answer this cognitive need and that the contribution to scientific systematization is not the sole but their *primary function*. This is also the reason for why he rejected instrumentalistic arguments based on Craig's theorem, see section 2.2. Nature's contribution to empirical irresolubility of "global

---

[39] Imagine a *grid-world* where positions are fully characterized by tuples (n,m) with $n, m \in \mathcal{Z}$. The observation sentences shall have the form $O(n,m)$ if the position (n,m) is "occupied" and $\overline{O}(n,m)$ otherwise. (As the formulation implies for no position $(O\&\overline{O})(n,m)$.) Suppose the total number of occupied positions is N. Every state-description of the grid-world then is "finite" in Quine's sense. If N is not finite there are two cases. The actual state can be summed up by a deterministic or recursive law, like for instance (T1): $O(n,m)$ iff n is even and m odd, and $\overline{O}(n,m)$ in all other cases including $(0,0)$. Alternatively, the actual state may either not admit a "mechanical", finite description in principle or none is found. The use of "theoretical concepts" save the day. An investigator for instance may postulate that there is a "propensity" $\text{prob}(n,m)$ for every point to be occupied and a positive field-strength $A(n,m)$ at every point, bounded away from 0, such that (T2): $\text{prob}(n,m) = \frac{1}{S}\exp\{-A(n,m)(n^2+m^2)\}$. S is a normalization factor. A theory like this may be the best one can do in giving a finite description of the grid-world.

science" is exhausted by forcing us to employ (finitely axiomatizable) theories for the purpose of prediction and control. At the same time – according to Quine – the thesis reflects what is "practicable feasible" for the human investigator of the material universe in two ways: her cognitive boundedness and her potential inability to recognize hidden equivalence when presented with what appears as an alternative global system of the world.

> [...] a last-ditch version of the thesis of under-determination would assert merely that our system of the world is bound to have empirically equivalent alternatives which, if we were to discover them, we would see no way of reconciling by reconstrual of predicates. (Quine, 1975, p. 327)

To show that two *prima facie* conflicting observationally indistinguishable alternatives are reinterpretable into each other by a systematic conversion requires *proof*. Proof sometimes remains elusive. Hence Quine's belief that it is *false* to hold that every pair of empirically equivalent alternatives is always recognizable as fully theoretically equivalent. The thesis asserts that circumstances can arise in which we remain *uncertain* if *prima facie* conflicting alternatives are fully, theoretically equivalent or not. Empirical irresolubility of global science and common sense is to be read as a claim about limits of human theorizing.[40] Interestingly, this interpretation of empirical irresolubility is compatible with robust realism with respect to the external world.

This analysis is unsatisfactory on two accounts. *First*, the origin of undetermination is only partially explained by our need to use concepts referring to non-observables to account for an infinite of data. After all, an unbounded amount of data could be compressed into a managable form, without a theoretical go-between, by way of recursive or a deterministic law, provided the world we are in "cooperates". So, it is rather

---

[40]Quine (1975 p. 326): "The more closely we examine the [under-determination – T.B.] thesis, the less we seem to be able to claim for it as a theoretical thesis; but it retains significance in terms of what is practicable feasible." This is the only place, to my knowledge, where Quine discusses the possibility of two readings of the thesis.

the logical structure of theory and the hypothetico-deductive nature of testing and confirming that gives rise to empirical irresolubility. Not theory *per se* is problematic in this respect, but the structure of theory. Moreover, the argument from the "utility" of theory seems to be open to the charge that it is overly *instrumentalist*. One looks in theories for epistemic virtues of explanatory depth, unificatory force, understanding and scope in themselves. Successful prediction and control are not merely "check-points", or primary or final utilities of theories. Rather, successful, novel prediction is the criterion for explanatory depth and scope. Quine's reply is, I think, that he was interested in the basics of theory formation – not in the epistemic value of theory. However, Quine's conclusion with regard to the likelihood of underdetermination of scientific theories depends on whether additional epistemic criteria are taken into account or not. *Second*, it cannot be right that the significance of the underdetemination thesis is as a reminder that we may fail in finding a way to show *prima facie* rival global theories to be compatible. There is a way to this end: Davidson's 'disambiguation' trick discussed in section 5.1. For that reason, in later writings Quine had shelved the pragmatic interpretation altogether and appeared to have returned to a straight-forward interpretation:

> What the empirical under-determination of global science shows is that there are various defensible ways of conceiving the world. (Quine, 1992a, p. 102)

I close this section by noting that the thesis of under-determination, or empirical irresolubility has *methodological import* for the naturalistic program in epistemology. If "reality can be conceived in many different ways" then it is part of the program to explain why humans came up with precisely that system of "global science", in both form and substance, that we find currently in textbooks and scientific journals. Considerations of *historical* contingencies and eventualities merely reflect on the rate and direction of scientific progress. And it seems wildly implausible that some chance event in the prehistory of the human species should have laid the conceptual foundations for modern physics as we know it. If indeed rival alternatives to our "best" system of beliefs exist, then it is not easy to account for why our "cognitive" species developed just the current system instead of a *radically* variant conception of reality, from a naturalist point of view.

Bibliography

Achinstein, P. (1994). Jean Perrin and Molecular Reality. *Perspectives on Science*, 2:396–427.

Ayer, A. J. (1982). *The Central Questions of Philosophy*. Harmondsworth: Penguin Books. Reprint of the 1973 edition.

Bain, J. (1999). Weinberg on QFT: Demonstrative Induction and Underdetermination. *Synthese*, 117:1–30.

Balzer, W., Moulines, C. U., and Sneed, J. S. (1987). *An Architectonic for Science*. Dordrecht, Holland: D. Reidel.

Barrett, R. B. and Gibson, R. F., editors (1990). *Perspectives on Quine*. Oxford; Basil Blackwell.

Bell, D. and Vossenkuhl, W., editors (1992). *Science and Subjectivity: The Vienna Circle and Twentieth Century Philosophy*. Berlin: Akademie Verlag.

Bergström, L. (1990). *Quine on Underdetermination*, chapter 3, pages 38–52. In Barrett and Gibson (1990).

Bohnert, H. (1968). In Defense of Ramsey's Elimination Method. *Journal of Philosophy*, 65:275–281.

Bonk, T. (1994). Why Has de Broglie's Theory Been Rejected? *Studies in the History and Philosophy of Science*, 25:375–396.

Bonk, T. (1997). Newtonian Gravity, Quantum Discontinuity and the Determination of Theory by Evidence. *Synthese*, 112:53–73.

Bonk, T. (2001a). How Reichenbach solved the Measurement Problem. *Dialectica*, 55:291–314.

Bonk, T. (2001b). *Scepticism under New Colors? Stroud's Criticism of Carnap*, chapter 4. In Bonk (2002).

Bonk, T., editor (2002). *Language, Truth, and Knowledge. Contributions to the Philosophy of Rudolf Carnap*. Dordrecht: Kluwer.

Bonk, T. (2004). Quine und der Realismus. *Zeitschrift für Philosophische Forschung*, 2:23–44.

Boyd, R. (1973). Realism, Underdetermination, and a Causal Theory of Evidence. *Nous*, VII:1–12.

Boyd, R. (1984). *The Current Status of Scientific Realism*, pages 41–82. In Leplin (1984).

Buck, R. and Cohen, R., editors (1971). *Boston Studies in the Philosophy of Science*, volume VIII. Dordrecht: Reidel.

Butts, R. E., editor (1968). *William Whewell's Theory of Scientific Method*. Pittsburgh, PA: University of Pittsburgh Press.

C. Batlle, J. Gomis, J. M. P. and Roman-Roy, N. (1986). Equivalence between the Lagrangian and Hamiltonian formalism for constrained systems. *Journal of Mathematical Physics*, 27:2953–2962.

Carnap, R. (1934). *Logische Syntax der Sprache*. Springer: Wien.

Carnap, R. (1955). *Meaning and synonymity in natural languages*, pages 233–247. In Carnap (1988). First published in *Philosophical Studies* 7 (1955) 33–47.

Carnap, R. (1961). *Der Logische Aufbau der Welt*. Hamburg: Felix Meiner, 2. edition.

Carnap, R. (1963a). *Carl G. Hempel on Scientific Theories*, pages 958–966. In Schilpp (1963).

Carnap, R. (1963b). *My views on Ontological Problems*, pages 868–873. In Schilpp (1963).

Carnap, R. (1963c). *Nelson Goodman on Der Logische Aufbau der Welt*, pages 944–947. In Schilpp (1963).

Carnap, R. (1988). *Meaning and Necessity*. Chicago, IL: University of Chicago Press. Reprint of the enlarged 1956 edition.

Chomsky, N. (1969). *Quine's empirical assumptions*, chapter 4, pages 53–68. In Davidson and Hintikka (1969).

Chomsky, N. (1975). *Reflections on Language*. New York: Pantheon Books.

Chomsky, N. (1980). *Rules and Representations*. New York: Columbia University Press.

Christensen, D. (1990). The Irrelevance of Bootstrapping. *Philosophy of Science*, 57:644–662.

Christopher, P. (1992). *A Study of Concepts*. Cambridge, MA: MIT Press.

Churchland, P. M. and Hooker, C. A., editors (1985). *Images of Science*. Chicago, IL: University of Chicago Press.

Clark, E. V. (1993). *The Lexicon in Acquisition*. Cambridge: Cambridge University Press.

Clark, P. and Hale, B., editors (1994). *Reading Putnam*. Oxford: Oxford University Press.

Cohen, R. S., editor (1976). *Essays in Memory of Imre Lakatos*. Dordrecht: Reidel.

Corcoran, J. (1983). *Alfred Tarski: Logic, Semantics, and Metamathematics. Papers from 1923 to 1938*. Indianapolis, IN: Hackett, 2. edition.

Craig, W. (1953). On Axiomatizability within a System. *Journal of Symbolic Logic*, 18:30–32.

Craig, W. (1956). Replacement of Auxiliary Expressions. *Philosophical Review*, 65:38–55.

Dam, H. V. and Wigner, E. P. (1966). Classical relativistic mechanics of interacting point particles. *Reviews of Modern Physics*, 136:1576–1582.

Davidson, D. (1977). *Reality without Reference*, chapter 15, pages 215–225. In Davidson (1984).

Davidson, D. (1984). *Inquiries into Truth and Interpretation*. Oxford: Claredon Press.

Davidson, D. and Hintikka, J., editors (1969). *Words and Objections: Essays on the Work of W. V. Quine*. Dordrecht: D. Reidel.

Descartes, R. (1956). *Die Prinzipien der Philosophie*. Hamburg: F. Meiner. Buchenau (trans.).

Deser, S. and Ford, K. W., editors (1965). *Lectures on General Relativity*. Englewood Cliffs, NJ: Prentice-Hall.

Devitt, M. (1997). *Realism and Truth*. Princeton: Princeton University Press, 2. edition.

Dorling, J. (1973). Demonstrative Induction: Its Significant Role in the History of Physics. *Philosophy of Science*, 40:360–372.

Douven, I. (1996). *Natural Kinds, Scientific Methodology, and Underdetermination*, pages 55–86. In Douven and Horsten (1996).

Douven, I. and Horsten, L., editors (1996). *Realism in the Sciences*. Leuven: Leuven University Press.

Duhem, P. (1978). *Ziel und Struktur wissenschaftlicher Theorien*. Hamburg: Felix Meiner.

Dummett, M. (1973). *The significance of Quine's Indeterminacy Thesis*, chapter 22, pages 375–419. In Dummett (1978).

Dummett, M. (1978). *Truth and other Enigmas*. Cambridge, MA: Harvard University Press.

Earman, J. S., editor (1977). *Foundations of Space-Time Theories*. Minneapolis, MN: University of Minnesota Press.

Earman, J. S., editor (1983). *Testing Scientific Theories*. Minneapolis, MN: University of Minnesota Press.

Earman, J. S., editor (1992). *Bayes or Bust? A Critical Examination of Bayesian Confirmation Theory*. Cambridge, MA: MIT Press.

Earman, J. S. (1993). Underdetermination, Realism, and Reason. *Midwest Studies in Philosophy*, 18:19–38.

Earman, J. S. and Glymour, C. (1988). What Revisions does Bootstrap Testing need? A Reply. *Philosophy of Science*, 55:260–264.

Einstein, A. and Infeld, L. (1938). *The Evolution of Physics*. New York: Simon & Schuster.

Ellis, B. (1985). *What Science Aims to Do?*, pages 48–74. In Churchland and Hooker (1985).

English, J. (1973). Underdetermination: Craig and Ramsey. *Journal of Philosophy*, 70:453–462.

Feyerabend, P. K. (1958). *Reichenbach's Interpretation of Quantum Mechanics*, chapter 15, pages 236–247. Volume I of Feyerabend (1981).

Feyerabend, P. K. (1981). *Realism, rationalism and scientific method. Philosophical papers*, volume I. Cambridge: Cambridge University Press.

Feyerabend, P. K. and Maxwell, G., editors (1966). *Mind, Matter and Method*. Minneapolis, MN: University of Minnesota Press.

Field, H. H. (1974). *Quine and Correspondence*, chapter 7, pages 199–221. In Field (2001).

Field, H. H. (1975). Conventionalism and Instrumentalism in semantics. *Nous*, 9:375–405.

Field, H. H. (1998). *Which Undecidable Sentences Have Determinate Truth Values?*, chapter 12, pages 332–360. In Field (2001).

Field, H. H. (2000). *Indeterminacy, Degree of Belief, and Excluded Middle*, chapter 10, pages 278–311. In Field (2001).

Field, H. H. (2001). *Truth and the Absence of Fact*. Oxford: Clarendon Press.

Fraenkel, A. (1928). *Einleitung in die Mengenlehre*. Berlin: Springer, 3. edition.

Friedman, M. (1975). Physicalism and the Indeterminacy of Translation. *Nous*, 9:353–373.

Friedman, M. (1983). *Foundations of Space-Time Theories*. Princeton, NJ: Princeton University Press.

Gaifman, H. (1975). Ontology and Conceptual Frameworks. *Erkenntnis*, 9:329–353.

Gardner, M. (1976). *The Unintelligibility of "Observational Equivalence"*, chapter 9, pages 104–116. In Suppe and Asquith (1976).

George, A. (1986). Whence and Whither the Debate Between Quine and Chomsky? *Journal of Philosophy*, 83:489–499.

George, A., editor (1989). *Reflections on Chomsky*. Cambridge: Blackwell.

Gibson, R. (1982). *The Philosophy of W.V. Quine*. Tampa, FL: University of South Florida.

Gibson, R. (1988). *Enlightened Empiricism*. Tampa, FL: University of South Florida.

Giere, R. (1993). Underdetermination, Relativism, and Perspectival Realism. Minnesota Center for Philosophy of Science.

Glymour, C. (1971). *Theoretical Realism and Theoretical Equivalence*. Volume VIII of Buck and Cohen (1971).

Glymour, C. (1980a). Discussion: Hypothetico-Deductivism is hopeless. *Philosophy of Science*, 47:322–325.

Glymour, C. (1980b). *Theory and Evidence*. Princeton, NJ: Princeton University Press.

Goldman, A. H. (1988). *Empirical Knowledge*. Berkeley: University of California Press.

Goodman, N. (1963). *The Significance of Der Logische Aufbau der Welt*, chapter 18. In Schilpp (1963).

Goodman, N. (1966). *The Structure of Appearance*. Indianapolis, IN: Bobbs-Merrill 2nd edition.

Goodman, N. (1978). *Ways of Worldmaking*. Hackett: Indianapolis.

Grünbaum, A. (1960). *The Duhemian Argument*, pages 116–131. In Harding (1976).

Grünbaum, A. and Salmon, W. C., editors (1988). *The Limitations of Deductivism*. Berkeley: University of California Press.

Haldane, E. S. and Ross, G. R. T., editors (1955). *The Philosophical Works of Descartes*. Dover.

Hallett, M. (1994). *Putnam and the Skolem Paradox*, chapter 4, pages 66–97. In Clark and Hale (1994).

Hanson, N. R. (1966). *Equivalence: The Paradox of Theoretical Analysis*, pages 413–430. In Feyerabend and Maxwell (1966).

Harding, S. G., editor (1976). *Can Theories be Refuted? Essays on the Duhem-Quine Thesis*. Dordrecht, Holland: D. Reidel.

Hegselmann, R., editor (1979). *Otto Neurath: Wissenschaftliche Weltauffassung, Sozialismus und Logischer Empirismus*. Frankfurt a.M.: Suhrkamp.

Hempel, C. G. (1951). *Empiricist Criteria of Cognitive Significance: Problems and Changes*, chapter 4. In Hempel (1966).

Hempel, C. G. (1958a). *Studies in the Logic of Confirmation*, chapter 1. In Hempel (1966).

Hempel, C. G. (1958b). *The Theoretician's Dilemma*, chapter 8. In Hempel (1966).

Hempel, C. G. (1966). *Aspects of Scientific Explanation*. New York: The Free Press.

Hermes, H. (1950). Zum Begriff der Axiomatisierbarkeit. *Mathematische Nachrichten*, IV:343–347.

Hilbert, D. and Ackermann, W. (1959). *Grundzüge der Theoretischen Logik*. Berlin: Springer, 4. edition. First edition 1928.

Hill, R. N. (1967). Instantaneous action-at-a-distance in classical relativistic mechanics. *Journal of Mathematical Physics*, 8:201–220.

Hilpinen, R., editor (1980). *Rationality in Science*. Boston, MA: D. Reidel.

Hintikka, J., editor (1975). *Rudolf Carnap: Logical Empiricist*. Dordrecht: D.Reidel.

Hodges, W. (1997). *A shorter model theory*. Cambridge: Cambridge University Press.

Hoffmann, B., editor (1966). *Perspectives in Geometry and Relativity*. Bloomington, IN: Indiana University Press.

Horwich, P. (1982). How to Choose Among Empirically Indistinguishable Theories. *Journal of Philosophy*, LXXIX:61–77.

Horwich, P. (1986). *A Defence of Conventionalism*, chapter 6, pages 163–188. In Macdonald and Wright (1986).

Horwich, P. (1991). On the Nature and Norms of Theoretical Commitment. *Philosophy of Science*, pages 1–14.

Howson, C. (2000). *Hume's Problem. Induction and the Justification of Belief.* Oxford: Clarendon Press.

Howson, C. and Urbach, P. (1989). *Scientific Reasoning: The Bayesian Approach.* La Salle, IL: Open Court.

Hume, D. (1902). *Enquiries Concerning the Human Understanding and Concerning the Principles of Morals.* Claredon Press: Oxford. Reprinted from the Posthumous Edition of 1777. L. A. Selby-Bigge (ed.) Impression of 1966.

J. C. C. McKinsey, A. C. S. and Suppes, P. (1953). Axiomatic Foundations of Classical Mechanics. *Journal of Rational Mechanics and Analysis*, 2:253–272.

Kitcher, P. (2001). Real Realism. *Philosophical Review*, 110:151–197.

Kukla, A. (1993). Laudan, Leplin, Empirical Equivalence and Underdetermination. *Analysis*, 53:1–7.

Kukla, A. (1994). Non-empirical Theoretical Values and the Argument from Underdetermination. *Erkenntnis*, 41:157–170.

Kukla, A. (1996). Does Every Theory Have Empirically Equivalent Rivals? *Erkenntnis*, 44:137–166.

Kukla, A. (1998). *Studies in Scientific Realism.* New York: Oxford University Press.

Lakoff, G. (1987). *Women, Fire, and Dangerous Things. What Categories reveal about the mind.* Chicago, IL: University of Chicago Press.

Laudan, L. (1965). Grünbaum on "The Duhemian Argument", pages 155–161. In Harding (1976).

Laudan, L. (1996). *Beyond Positivism and Relativism. Theory, Method and Evidence.* Boulder, CO.: Westview Press.

Laymon, R. (1994). Demonstrative Induction, Old and New Evidence and the Accuracy of the Electrostatic Inverse Square Law. *Synthese*, 99:23–58.

Leeds, S. (1973). How to Think about Reference. *Journal of Philosophy*, LXX:485–503.

Leplin, J., editor (1984). *Scientific Realism.* Berkeley: University of California Press.

Leplin, J. (1997a). *A Novel Defense of Realism.* Oxford: Oxford University Press.

Leplin, J. (1997b). The Underdetermination of Total Theories. *Erkenntnis*, 47:203–215.

Leplin, J. (2000). The epistemic status of auxiliary hypotheses: A Reply to Douven. *The Philosophical Quarterly*, 50:376–380.

Leplin, J. and Laudan, L. (1993). Determination Undeterred: Reply to Kukla. *Analysis*, 53:8–16.

LePore, E., editor (1986). *Truth and Interpretation.* Oxford: Basil Blackwell.

Lewis, D. K. (1974). *Radical Interpretation*, chapter 8, pages 108–118. In Lewis (1983).

Lewis, D. K. (1983). *Philosophical Papers.* Oxford, UK: Oxford University Press.

Lewis, D. K. (1984). Putnam's Paradox. *Australasian Journal of Philosophy*, 62:214–236.

Longino, H. E. (2002). *The Fate of Knowledge*. Princeton: Princeton University Press.

Macdonald, G. and Wright, C., editors (1986). *Fact, Science, and Morality*. Oxford: Basil Blackwell.

Mackey, G. W. (1963). *Mathematical Foundations of Quantum Mechanics*. New York: Benjamin.

Malament, D. (1977). *Observationally Indistinguishable Space-times*, pages 61–80. In Earman (1977).

Massey, G. J. (1978). *Indeterminacy, Inscrutability, and Ontological Relativity*, chapter 3. In Rescher (1978).

Maxwell, J. C. (1958). *Collected Scientific Papers*. New York: Dover.

Mayo, D. G. (1996). *Error and the Growth of Experimental Knowledge*. Chicago, IL: University of Chicago Press.

McCormick, P. J. (1996). *Starmaking: Realism, Anti-realism, and Irrealism*. Boston, MA: MIT Press.

Mehlberg, H. (1958). *The Reach of Science*. Toronto: University of Toronto Press.

Mill, J. S. (1916). *A System of Logic Ratiocinative and Inductive*. London: Longmans, Green, and Co.

Miller, R. W. (1987). *Fact and Method*. Princeton, NJ: Princeton University Press.

Misner, C. W., Thorne, K. S., and Wheeler, J. A. (1973). *Gravitation*. San Francisco, CA: W. H. Freeman.

Moulines, C. U. *Ontologie, Reduktionismus, Einheit der Wissenschaft*. Unpublished Manuskript.

Moulines, C. U. (1997). The Concept of Universe from a Metatheoretical Point of View. *Poznan Studies*, 61:359–379.

Mundy, B. (1989). DIstant Action in Classical Electromagnetic Theory. *The British Journal for the Philosophy of Science*, 40:39–68.

Myhill, J. R. (1951). *On the ontological significance of the Löwenheim-Skolem theorem.* In White (1951).

Neurath, O. (1934). *Radikaler Physikalismus und 'Wirkliche Welt'*, chapter 2. In Hegselmann (1979).

Newman, M. H. A. (1928). Mr. Russell's "Causal theory of perception". *Mind*, XXXVIL:137–148.

Newton-Smith, W. (1978). *The underdetermination of theory by data.* In Hilpinen (1980).

Norton, J. D. (1993). The Determination of Theory by Evidence: The Case for Quantum Discontinuity, 1900–1915. *Synthese*, 97:1–31.

Norton, J. D. (1994). Science and Certainty. *Synthese*, 99:3–22.

Okasha, S. (1997). Laudan and Leplin on empirical equivalence. *British Journal for the Philosophy of Science*, 48:251–256.

Oreskes, N., Shrader-Frechette, K., and Belitz, K. (1994). Verification, Validation, and Confirmation of Numerical Models in the Earth Sciences. *Science*, 263:641–646.

P. A. French, T. E. U. and Wettstein, H. K., editors (1979). *Contemporary Perspectives in the Philosophy of Language*. Minneapolis: University of Minnesota Press.

Poincaré, H. (1901). *Electricité et Optique. Leçons Professées a la Sorbonne*. Paris: Gauthiers-Villars. 2nd edition.

Poincaré, H. (1952). *Science and Hypothesis*. New York: Dover.

Putnam, H. (1957). *Three-valued Logic*, chapter 9. Volume I of Putnam (1979).

Putnam, H. (1965). *Craig's Theorem*, chapter 14, pages 228–236. Volume 3 of Putnam (1983b).

Putnam, H. (1972). *Other Minds*, chapter 17, pages 342–361. Volume 2 of Putnam (1975a).

Putnam, H. (1975a). *Mind, Language and Reality. Philosophical Papers*, volume 2. New York: Cambridge University Press.

Putnam, H. (1975b). *The Refutation of Conventionalism*, chapter 9, pages 153–191. Volume 2 of Putnam (1975a).

Putnam, H. (1977). *Models and Reality*, chapter 1, pages 1–25. Volume 3 of Putnam (1983b).

Putnam, H. (1978). *Meaning and the Moral Sciences*. London: Routledge.

Putnam, H. (1979). *Mathematics, Matter and Method. Philosophical Papers*, volume I. Cambridge: Cambridge University Press.

Putnam, H. (1981). *Reason, Truth and History*. Cambridge: Cambridge University Press.

Putnam, H. (1983a). *Equivalence*, chapter 2, pages 26–45. Volume 3 of Putnam (1983b).

Putnam, H. (1983b). *Realism and Reason. Philosophical Papers*, volume 3. Cambridge: Cambridge University Press.

Putnam, H. (1983c). *Vagueness and Alternative Logic*, chapter 15, pages 271–286. Volume 3 of Putnam (1983b).

Putnam, H. (1987). *The many faces of Realism. The Paul Carus Lectures*. La Salle, IL.: Open Court.

Putnam, H. (1989). *Model theory and the 'factuality' of semantics*, chapter 11, pages 213–232. Volume 83 of George (1986).

Quine, W. V. (1960a). *Posits and Reality*, chapter 20, pages 233–241. In Quine (1966).

Quine, W. V. (1960b). *Word and Object*. Cambridge, MA: MIT Press.

Quine, W. V. (1962). *A Comment on Grünbaum's Claim*, page 132. In Harding (1976).

Quine, W. V. (1963). *On Simple Theories in a Complex World*, chapter 21, pages 242–245. In Quine (1966).

Quine, W. V. (1964). *Ontological Reduction*, chapter 17, pages 199–207. In Quine (1966).

Quine, W. V. (1966). *The Ways of Paradox*. New York: Random House.

Quine, W. V. (1968). *Epistemology Naturalized*, chapter 3. In Quine (1969c).

Quine, W. V. (1969a). *Existence and Quantification*, chapter 4, pages 91–113. In Quine (1969c).

Quine, W. V. (1969b). *Ontological Relativity*, chapter 2, pages 26–68. In Quine (1969c).

Quine, W. V. (1969c). *Ontological Relativity and Other Essays*. New York: Columbia University Press.

Quine, W. V. (1969d). *Reply to Chomsky*, pages 302–311. In Davidson and Hintikka (1969).

Quine, W. V. (1970). On the Reasons for the Indeterminacy of Translation. *Journal of Philosophy*, 67:178–183.

Quine, W. V. (1973a). *The Limits of Knowledge*, chapter 9, pages 59–67. In Quine (1966).

Quine, W. V. (1973b). *The Roots of Reference*. La Salle, IL: Open Court.

Quine, W. V. (1975). On Empirically Equivalent Systems of the World. *Erkenntnis*, 9:313–328.

Quine, W. V. (1976). *Whither Physical Objects?*, pages 497–504. In Cohen (1976).

Quine, W. V. (1977). *Facts of the Matter*, pages 176–196. In Shahan and Merrill (1977).

Quine, W. V. (1980). *Things and Their Place in Theories*, chapter 1, pages 1–23. In Quine (1981).

Quine, W. V. (1981). *Theories and Things*. Cambridge, MA: Belknap Press.

Quine, W. V. (1987). Indeterminacy of Translation Again. *Journal of Philosophy*, LXXXIV:5–10.

Quine, W. V. (1988). *Comment on Bergström*, chapter 3, pages 53–54. In Davidson and Hintikka (1969).

Quine, W. V. (1990). *Three Indeterminacies*, chapter 1, pages 1–16. In Barrett and Gibson (1990).

Quine, W. V. (1991). Two Dogmas in Retrospect. *Canadian Journal of Philosophy*, 21:265–274.

Quine, W. V. (1992a). *Pursuit of Truth*. Cambridge, MA: Harvard University Press. Revised edition.

Quine, W. V. (1992b). Structure and Nature. *Journal of Philosophy*, 89:5–9.

Quine, W. V. (1995). *From Stimulus to Science*. Cambridge, MA: Harvard University Press.

Reichenbach, H. (1938). *Experience and Prediction.* Chicago, IL: University of Chicago Press. Fourth Impression 1952.

Reichenbach, H. (1958). *The Philosophy of Space and Time.* New York: Dover. Translation of Philosophie der Raum-Zeit-Lehre.

Rescher, N., editor (1978). *Studies in Ontology.* American Philosophical Quarterly Monograph Series. Oxford: Basil Blackwell.

Rescher, N. (1989). *Cognitive Economy. The Economic Dimension of the Theory of Knowledge.* Pittsburgh, PA: University of Pittsburgh Press.

Rozeboom, W. W. (1960). Studies in the Empiricist Theory of Scientific Meaning. Part II: On the Equivalence of Scientific Theories. *Philosophy of Science*, 27:366 – 373.

Russell, B. (1921). *Analysis of Mind.* London: George Allen & Unwin.

SantAnna, A. (1996). An axiomatic framework for classical particle mechanics without force. *Philosophia Naturalis*, 33:187–203.

Scheck, F. (1990). *Mechanics.* Berlin: Springer, 2. edition.

Scheffler, I. (1968). Reflections on the Ramsey Method. *Journal of Philosophy*, 65:269–274.

Scheffler, I. (1996). *The Wonderful Worlds of Goodman*, chapter 8, pages 133–141. In McCormick (1996).

Schilpp, P. A., editor (1963). *The Philosophy of Rudolf Carnap.* La Salle, Ill., London: Open Court.

Schilpp, P. A. and Hahn, L. E., editors (1986). *The Philosophy of W. V. Quine.* La Salle, Ill., London: Open Court.

Schlick, M. (1925). *Allgemeine Erkenntnislehre.* Wien: Springer, 2. edition.

Searle, J. R. (1969). *Speach Acts. An Essay in the Philosophy of Language.* Cambridge: Cambridge University Press. 1996 reprint.

Sellars, W. (1963). *Science, Perception and Reality.* Atascadero, CA: Ridgeview Publishing Company. 1991 reprint.

Shahan, R. W. and Merrill, K. R., editors (1977). *American Philosophy from Edwards to Quine.* University of Oklahoma: University of Oklahoma Press.

Shapiro, S. (1997). *Philosophy of Mathematics.* Oxford: Oxford University Press.

Shoemaker, S. (1969). Time without Change. *Journal of Philosophy*, 66:363–381.

Sober, E. (1996). Parsimony and Predictive Equivalence. *Erkenntnis*, 46:167–197.

Solomon, M. (1989). Quine's point of view. *Journal of Philosophy*, LXXXVI:113–136.

Spelke, E. S., Breinlinger, K., Macomber, J., and Jacobson, K. (1992). Origins of Knowledge. *Psychological Review*, 99:605–632.

Stanford, P. K. (2001). Refusing the Devil's Bargain: What Kind of Underdetermination Should we take Seriously? *Philosophy of Science*, 68:S1–S12.

Stegmüller, W. (1973). *Struktur und Dynamik wissenschaftlicher Theorien.* Springer.

Strawson, P. F. (1966). *The Bounds of Sense.* London: Routledge.

Stroud, B. (1968–1969). Conventionalism and the Indeterminacy of Translation. *Synthese*, 19:82–96.

Stroud, B. (1984). *The Significance of Philosophical Scepticism.* Oxford: Clarendon Press.

Suppe, F. and Asquith, P. D., editors (1976). *Proceedings of the 1976 Biennial Meeting of the Philosophy of Science Association*. Ann Arbor, MI: Edwards Brothers.

Swinburne, R. (1968). *Space and Time*. London: Macmillan.

Tarski, A. and Lindenbaum, A. (1934–1935). *On the Limitations of the Means of Expression of Deductive Theories*, chapter XIII, pages 384–392. In Corcoran (1983), 2. edition. Published as: Über die Beschränktheit der Ausdrucksmittel deduktiver Theorien, in: *Ergebnisse eines Mathematischen Kolloquiums*, fascicule 7, pp. 15–22.

Thirring, W. (1961). An Alternative Approach to the Theory of Gravitation. *Annals of Physics*, 16:96–117.

Trautmann, A. (1964). *Comparsion of Newtonian and Relativistic Theories of Space-Time*, chapter 42, pages 413–425. In Hoffmann (1966).

Trautmann, A. (1965). *Foundations and Current Problems of General Relativity*. In Deser and Ford (1965).

van Fraassen, B. (1980a). *The Scientific Image*. Oxford: Clarendon Press.

van Fraassen, B. (1980b). *Theory Comparison and Relevant Evidence*, chapter I, pages 27–42. In Earman (1983).

Wallace, J. (1979a). *Only in the Context of Sentence do Words have Meaning*, pages 305–325. Minneapolis, MN: University of Minnesota Press.

Wallace, J. (1979b). *Translation Theories and the Decipherment of Linear B*, chapter 11. In LePore (1986).

Weyl, H. (1948). *Philosophie der Mathematik und Naturwissenschaft*. München: Leibniz Verlag, 2. edition. Neu herausgegeben von M. Schröder.

Wheeler, J. A. and Feynman, R. P. (1949). Classical Electrodynamics in Terms of Direct Interparticle Action. *Reviews of Modern Physics*, 21:425–434.

Whewell, W. (1849). *Mr. Mill's Logic.* In Butts (1968).

White, M., editor (1951). *Academic Freedom, Logic, and Religion.* Pennsylvania: University of Pennsylvania Press.

Whitehead, A. N. (1955). *An Enquiry Concerning the Principles of Natural Knowledge.* Cambridge: At the University Press. Reprint of the second edition.

Wilson, M. (1980). The Observational Uniqueness of Some Theories. *Journal of Philosophy*, 77:208–233.

Wilson, M. (1981). The Double Standard in Ontology. *Philosophical Studies*, 39:409–427.

Winnie, J. (1967). The Implicit Definition of Theoretical Terms. *British Journal of Philosophy of Science*, 18:223–229.

Winnie, J. (1975). *Theoretical Analycity.* In Hintikka (1975).

Wittgenstein, L. (1958). *Philosophical Investigations. Part I.* Oxford: Blackwell.

Wittgenstein, L. (1989). *Logisch-philosophische Abhandlung.* Frankfurt a. M.: Suhrkamp. Published 1921 in *Annalen der Natur- und Kulturphilosophie*, Vol. 14. pp. 185–262. Critical edition by B. McGuiness und J. Schulte.

Wright, C. (1986). *Scientific Realism, Observation, and the Verification Principle*, chapter 9. In Macdonald and Wright (1986).

Wright, C. (1992). *Scientific Realism and Observation Statements*, pages 21–46. In Bell and Vossenkuhl (1992).

Index

auxiliary hypothesis, 3, 92, 98, 100, 101, 105, 106
Ayer, A.J., 40

Bohr, N., 118
Boyd, R., 12, 129, 130
    background rule, 131
    underdetermination, 2

Carnap, R., 41, 47, 100, 142, 215, 236
    coordinate language, 148
    Duhem's thesis, 105
    inductive linguist, 231, 233, 236
    intension, 214, 232
    ostensive reference, 231
causality, 181
Chomsky, N.
    indeterminacy, 220
confirmation
    analogies, 127, 129
    converse consequence condition, 93, 128
    demonstrative induction, 119, 122
    holism, 97
    hypothetico-deductive, 91, 93, 99, 116
    instance, 112, 114
    novelty, 95
    special consequence condition, 128
    total evidence condition, 91
constructivism, 141, 147
content, 170, 203
    auxiliary hypotheses, 92
    falsity, 18, 92
    observational, 92
    Quine, W.V., on, 108
    relative content, 92, 100
    truth, 18, 92
conventionalism, 29
    coordinative definition, 29, 42, 183
    global, 195
Copernicus, N., 5
Craig, W., 52, 136

Davidson, D., 179, 254
Descartes, R., 6
    refraction, 90
    underdetermination, 6, 14
descriptivism, 178, 191
Devitt, M., 13
Douven, I., 33
duality, 23
Duhem, P., 104, 120

Duhem-Quine thesis, 94, 98, 105, 119, 252
Dummett, M., 13

Earman, J., 10, 37
economy
    conceptual, 199
Einstein, A., 7
electro-dynamics
    classical, 79
    Wheeler-Feynman, 79, 166
equivalence
    definitional, 47
    hd-equivalence, 92
    logical, 46
    mathematical realism, 167
    no common model, 51
    observational, 111
    physical, 69
    reduction of, 134
    relativized, 46

Feynman, R., 79
fictionalism, 201
    limit concepts, 201
Field, H., 43
    continuum hypothesis, 251
    degrees of belief, 240, 243
    indeterminacy, 238, 241
    non-classical logic, 242
    partial denotation, 238
    vagueness, 240
finite, 159
force, 143
    action-at-a-distance, 79, 81
    differential, 83
    elimination of, 153
    universal, 82, 83, 130, 136
    universal gravitation, 121
Fowler, R., 120
Fraenkel, A., 58, 59
Frank, Ph., 36
Friedman, M., 29, 37, 249
    on unifying power, 135
    unification rule, 136

gauge transformation, 73
geometry, 82, 83
    Euclidean, 61, 62, 163
Giere, R., 13, 34
Glymour, C., 51, 112, 113, 116
Goodman, N., 35, 37, 161
    point vs line, 164, 165
Grünbaum, A., 105

Hamiltonian mechanics, 74
    equivalence with Lagrange mechanics, 75
Hanson, N.R., 47
Hare, R.M.
    objective values, 9, 40
Heisenberg, W., 131
Hempel, C.G., 93, 112, 128, 168
Hermes, H., 52
Hertz, H., 255
Hilbert, D., 201
holism, 107
Horwich, P., 12, 20, 48, 57, 166, 192, 194
Hume, D., 34
Huyghens, C., 90

identity criteria, 47

indeterminacy, 7
  confirmational, 94, 98
  of beliefs, 229
  of translation, 210
  of truth conditions, 229
  semantical, 246
indispensability, 202, 205
indistinguishability
  disambiguation, 178, 179
  inverted spectrum, 209
  lawhood, 9
  neuromolecular, 208, 211
  observational, 8, 26, 64, 246
  ostensive, 234
  probability distributions, 92
  rational, 8
induction, 34, 37
  eliminative, 11
  simplicity, 34
instrumentalism, 12
inter-theory relation, 128, 129, 132
interdefinability, 151
irresolubility
  Turing, 41
isomorphism, 49, 50

Kant, I., 158
  antinomy of pure reason, 196
Kepler, J., 5
  Kepler's laws, 121, 128
Kitcher, Ph., 17, 20
Konstitutionssystem, 142
Kukla, A., 31, 169, 170

Löwenheim, L., 156
Lagrangian mechanics, 69, 70, 74, 76

Laudan, L., 5, 11, 105, 126–128, 169
Leplin, J., 5, 100, 126, 169
  content, 93
Lewis, D.
  indeterminacy, 229
  translation, 229
Longino, H., 13
Lorentz, H.A.
  electron theory, 101

Mach, E., 79
Mayo, D., 10
Mehlberg, H., 59, 62
methodology, 24
  demonstrative induction, 119
  falsificationism, 103
  hypothetico-deductive, 90, 97
Mill, J.S., 34
Miller, R.W., 31, 132, 181
mind, 14, 188

naturalism, 26, 253, 254
Norton, J., 119

observation
  and stimulus, 107
  observation categoricals, 108
  occasion sentence, 107
Ockham's razor, 199

particle physics
  standard model, 119
Peacocke, C., 204
Peirce, C.S., 23
Planck, M., 120
Poincaré, H., 193

disk, 193, 198
multiple mechanical models, 78
principle
　of bivalence, 43
Putnam, H., 34, 48, 65, 82, 151, 161, 172, 195
　indeterminacy, 221
　reference, 235
　vagueness, 245

qualia, 144, 208
quantum theory, 118, 131, 132, 135
　Pauli exclusion principle, 101
Quine, W.V., 1, 21, 39, 48, 68, 102, 211, 213, 228
　analytical hypotheses, 211, 212, 230
　behaviorism, 224
　continuum hypothesis, 251
　Craig's theorem, 54
　Duhem-Quine thesis, 104, 252
　egalitarian solution, 193
　empirical content, 108
　hypothetico-deductive methodology, 257
　indeterminacy, 215, 217, 219, 220, 229
　innate knowledge, 224, 227
　inverted spectrum, 210
　ostension, 233
　physicalism, 225
　proxy-function, 150, 236
　Ramsey sentence, 235
　reference, 235

second intension argument, 216
stimulus, 235
structuralism, 27
tandem solution, 193
translation, 220
translation manual, 159, 219
underdetermination, 1, 107, 217, 254–257

Ramsey sentence, 55, 57, 191
Ramsey, F.P., 55
　lawhood, 9
realism, 13, 15
　and naturalism, 26
　epistemological component, 17
　external world, 15
　hypothetico-deductive, 26, 181
　metaphysical, 26
　no miracle, 21, 132
　past events, 15
　perspectival, 13
　Quine on, 26
　scientific, 12, 16
　sentence meanings, 15
　structural, 18
reconstrual of predicates, 48
Reichenbach, H., 41, 82, 169, 181, 208, 246
　causality, 181
　coordinative definition, 183
　ego-centric language, 182
Russell, B., 15
　sentence meaning, 214

Sellars, W., 26
sentence
    isolated, 167
simplicity, 7, 16, 24, 118
    algorithmic, 36
    descriptive, 36
    inductive, 36
    parametric, 36
skepticism, 13, 30
    vat-operator, 194, 195
Sober, E., 16
    error distribution, 92
    other minds, 208
space of theories, 31
Stanford, P. K., 32
Stroud, B., 253
    stimulus-observability, 254
    translation, 218
structuralism, 167
Suppes, P., 79
symmetry, 23, 68, 78

Tarski, A., 60, 61, 236
tempered pluralism, 17
theorem
    Bertrand's, 121
    Craig's, 52
    Glymour's, 51
    Löwenheim-Skolem, 67, 156, 157
    Phytagoreisation, 64
    Putnam's, 65
    Winnie's, 63

underdetermination, 54, 57, 84, 119, 126, 215, 220

algorithms, 167, 175
all possible observations, 107
color qualia, 208
conceiving the world, 257
elimination, 40
global science, 21
Goodman's riddle, 37
Humean, 4
isolated sentence, 167
methodological, 4
observer-dependence, 169
ontological reduction, 162
other minds, 207
parameter dependence, 172
pragmatic interpretation, 256
Quine on, 1, 22, 103, 255, 256
strong, 1, 4, 34, 116
topology of time, 16
transient, 4
truth, 25
unification, 135
    unique to physics, 30, 132
weak, 4, 34, 231

variant, 46
    notational, 45, 47, 191
verificationism, 40

Wallace, D.
    decipherment vs. translation, 226
Weyl, H., 59
    holism, 105
    isomorphism, 59
    on length doubling, 203
Wheeler, J., 79

Whitehead, A.N.
    elimination of points, 161
Wigner, E., 81
Wilson, M., 110, 152
Winnie, J., 60, 63–65
Wittgenstein, L., 47, 233
Wright, C., 13

Boston Studies in the Philosophy of Science

Editor: Robert S. Cohen, *Boston University*

1. M.W. Wartofsky (ed.): *Proceedings of the Boston Colloquium for the Philosophy of Science, 1961/1962.* [Synthese Library 6] 1963 ISBN 90-277-0021-4
2. R.S. Cohen and M.W. Wartofsky (eds.): *Proceedings of the Boston Colloquium for the Philosophy of Science, 1962/1964.* In Honor of P. Frank. [Synthese Library 10] 1965 ISBN 90-277-9004-0
3. R.S. Cohen and M.W. Wartofsky (eds.): *Proceedings of the Boston Colloquium for the Philosophy of Science, 1964/1966.* In Memory of Norwood Russell Hanson. [Synthese Library 14] 1967 ISBN 90-277-0013-3
4. R.S. Cohen and M.W. Wartofsky (eds.): *Proceedings of the Boston Colloquium for the Philosophy of Science, 1966/1968.* [Synthese Library 18] 1969
ISBN 90-277-0014-1
5. R.S. Cohen and M.W. Wartofsky (eds.): *Proceedings of the Boston Colloquium for the Philosophy of Science, 1966/1968.* [Synthese Library 19] 1969
ISBN 90-277-0015-X
6. R.S. Cohen and R.J. Seeger (eds.): *Ernst Mach, Physicist and Philosopher.* [Synthese Library 27] 1970 ISBN 90-277-0016-8
7. M. Čapek: *Bergson and Modern Physics.* A Reinterpretation and Re-evaluation. [Synthese Library 37] 1971 ISBN 90-277-0186-5
8. R.C. Buck and R.S. Cohen (eds.): *PSA 1970.* Proceedings of the 2nd Biennial Meeting of the Philosophy and Science Association (Boston, Fall 1970). In Memory of Rudolf Carnap. [Synthese Library 39] 1971
ISBN 90-277-0187-3; Pb 90-277-0309-4
9. A.A. Zinov'ev: *Foundations of the Logical Theory of Scientific Knowledge (Complex Logic).* Translated from Russian. Revised and enlarged English Edition, with an Appendix by G.A. Smirnov, E.A. Sidorenko, A.M. Fedina and L.A. Bobrova. [Synthese Library 46] 1973 ISBN 90-277-0193-8; Pb 90-277-0324-8
10. L. Tondl: *Scientific Procedures.* A Contribution Concerning the Methodological Problems of Scientific Concepts and Scientific Explanation. Translated from Czech. [Synthese Library 47] 1973 ISBN 90-277-0147-4; Pb 90-277-0323-X
11. R.J. Seeger and R.S. Cohen (eds.): *Philosophical Foundations of Science.* Proceedings of Section L, 1969, American Association for the Advancement of Science. [Synthese Library 58] 1974 ISBN 90-277-0390-6; Pb 90-277-0376-0
12. A. Grünbaum: *Philosophical Problems of Space and Times.* 2nd enlarged ed. [Synthese Library 55] 1973 ISBN 90-277-0357-4; Pb 90-277-0358-2
13. R.S. Cohen and M.W. Wartofsky (eds.): *Logical and Epistemological Studies in Contemporary Physics.* Proceedings of the Boston Colloquium for the Philosophy of Science, 1969/72, Part I. [Synthese Library 59] 1974
ISBN 90-277-0391-4; Pb 90-277-0377-9
14. R.S. Cohen and M.W. Wartofsky (eds.): *Methodological and Historical Essays in the Natural and Social Sciences.* Proceedings of the Boston Colloquium for the Philosophy of Science, 1969/72, Part II. [Synthese Library 60] 1974
ISBN 90-277-0392-2; Pb 90-277-0378-7
15. R.S. Cohen, J.J. Stachel and M.W. Wartofsky (eds.): *For Dirk Struik.* Scientific, Historical and Political Essays in Honor of Dirk J. Struik. [Synthese Library 61] 1974 ISBN 90-277-0393-0; Pb 90-277-0379-5
16. N. Geschwind: *Selected Papers on Language and the Brains.* [Synthese Library 68] 1974 ISBN 90-277-0262-4; Pb 90-277-0263-2

Boston Studies in the Philosophy of Science

17. B.G. Kuznetsov: *Reason and Being*. Translated from Russian. Edited by C.R. Fawcett and R.S. Cohen. 1987   ISBN 90-277-2181-5
18. P. Mittelstaedt: *Philosophical Problems of Modern Physics*. Translated from the revised 4th German edition by W. Riemer and edited by R.S. Cohen. [Synthese Library 95] 1976   ISBN 90-277-0285-3; Pb 90-277-0506-2
19. H. Mehlberg: *Time, Causality, and the Quantum Theory*. Studies in the Philosophy of Science. Vol. I: *Essay on the Causal Theory of Time*. Vol. II: *Time in a Quantized Universe*. Translated from French. Edited by R.S. Cohen. 1980
    Vol. I: ISBN 90-277-0721-9; Pb 90-277-1074-0
    Vol. II: ISBN 90-277-1075-9; Pb 90-277-1076-7
20. K.F. Schaffner and R.S. Cohen (eds.): *PSA 1972*. Proceedings of the 3rd Biennial Meeting of the Philosophy of Science Association (Lansing, Michigan, Fall 1972). [Synthese Library 64] 1974   ISBN 90-277-0408-2; Pb 90-277-0409-0
21. R.S. Cohen and J.J. Stachel (eds.): *Selected Papers of Léon Rosenfeld*. [Synthese Library 100] 1979   ISBN 90-277-0651-4; Pb 90-277-0652-2
22. M. Čapek (ed.): *The Concepts of Space and Time*. Their Structure and Their Development. [Synthese Library 74] 1976   ISBN 90-277-0355-8; Pb 90-277-0375-2
23. M. Grene: *The Understanding of Nature*. Essays in the Philosophy of Biology. [Synthese Library 66] 1974   ISBN 90-277-0462-7; Pb 90-277-0463-5
24. D. Ihde: *Technics and Praxis*. A Philosophy of Technology. [Synthese Library 130] 1979   ISBN 90-277-0953-X; Pb 90-277-0954-8
25. J. Hintikka and U. Remes: *The Method of Analysis*. Its Geometrical Origin and Its General Significance. [Synthese Library 75] 1974
    ISBN 90-277-0532-1; Pb 90-277-0543-7
26. J.E. Murdoch and E.D. Sylla (eds.): *The Cultural Context of Medieval Learning*. Proceedings of the First International Colloquium on Philosophy, Science, and Theology in the Middle Ages, 1973. [Synthese Library 76] 1975
    ISBN 90-277-0560-7; Pb 90-277-0587-9
27. M. Grene and E. Mendelsohn (eds.): *Topics in the Philosophy of Biology*. [Synthese Library 84] 1976   ISBN 90-277-0595-X; Pb 90-277-0596-8
28. J. Agassi: *Science in Flux*. [Synthese Library 80] 1975
    ISBN 90-277-0584-4; Pb 90-277-0612-3
29. J.J. Wiatr (ed.): *Polish Essays in the Methodology of the Social Sciences*. [Synthese Library 131] 1979   ISBN 90-277-0723-5; Pb 90-277-0956-4
30. P. Janich: *Protophysics of Time*. Constructive Foundation and History of Time Measurement. Translated from German. 1985   ISBN 90-277-0724-3
31. R.S. Cohen and M.W. Wartofsky (eds.): *Language, Logic, and Method*. 1983
    ISBN 90-277-0725-1
32. R.S. Cohen, C.A. Hooker, A.C. Michalos and J.W. van Evra (eds.): *PSA 1974*. Proceedings of the 4th Biennial Meeting of the Philosophy of Science Association. [Synthese Library 101] 1976   ISBN 90-277-0647-6; Pb 90-277-0648-4
33. G. Holton and W.A. Blanpied (eds.): *Science and Its Public*. The Changing Relationship. [Synthese Library 96] 1976
    ISBN 90-277-0657-3; Pb 90-277-0658-1
34. M.D. Grmek, R.S. Cohen and G. Cimino (eds.): *On Scientific Discovery*. The 1977 Erice Lectures. 1981   ISBN 90-277-1122-4; Pb 90-277-1123-2

Boston Studies in the Philosophy of Science

35. S. Amsterdamski: *Between Experience and Metaphysics*. Philosophical Problems of the Evolution of Science. Translated from Polish. [Synthese Library 77] 1975
ISBN 90-277-0568-2; Pb 90-277-0580-1
36. M. Marković and G. Petrović (eds.): *Praxis*. Yugoslav Essays in the Philosophy and Methodology of the Social Sciences. [Synthese Library 134] 1979
ISBN 90-277-0727-8; Pb 90-277-0968-8
37. H. von Helmholtz: *Epistemological Writings*. The Paul Hertz / Moritz Schlick Centenary Edition of 1921. Translated from German by M.F. Lowe. Edited with an Introduction and Bibliography by R.S. Cohen and Y. Elkana. [Synthese Library 79] 1977
ISBN 90-277-0290-X; Pb 90-277-0582-8
38. R.M. Martin: *Pragmatics, Truth and Language*. 1979
ISBN 90-277-0992-0; Pb 90-277-0993-9
39. R.S. Cohen, P.K. Feyerabend and M.W. Wartofsky (eds.): *Essays in Memory of Imre Lakatos*. [Synthese Library 99] 1976
ISBN 90-277-0654-9; Pb 90-277-0655-7
40. Not published.
41. Not published.
42. H.R. Maturana and F.J. Varela: *Autopoiesis and Cognition*. The Realization of the Living. With a Preface to "Autopoiesis" by S. Beer. 1980
ISBN 90-277-1015-5; Pb 90-277-1016-3
43. A. Kasher (ed.): *Language in Focus: Foundations, Methods and Systems*. Essays in Memory of Yehoshua Bar-Hillel. [Synthese Library 89] 1976
ISBN 90-277-0644-1; Pb 90-277-0645-X
44. T.D. Thao: *Investigations into the Origin of Language and Consciousness*. 1984
ISBN 90-277-0827-4
45. F.G.-I. Nagasaka (ed.): *Japanese Studies in the Philosophy of Science*. 1997
ISBN 0-7923-4781-1
46. P.L. Kapitza: *Experiment, Theory, Practice*. Articles and Addresses. Edited by R.S. Cohen. 1980   ISBN 90-277-1061-9; Pb 90-277-1062-7
47. M.L. Dalla Chiara (ed.): *Italian Studies in the Philosophy of Science*. 1981
ISBN 90-277-0735-9; Pb 90-277-1073-2
48. M.W. Wartofsky: *Models*. Representation and the Scientific Understanding. [Synthese Library 129] 1979   ISBN 90-277-0736-7; Pb 90-277-0947-5
49. T.D. Thao: *Phenomenology and Dialectical Materialism*. Edited by R.S. Cohen. 1986
ISBN 90-277-0737-5
50. Y. Fried and J. Agassi: *Paranoia*. A Study in Diagnosis. [Synthese Library 102] 1976   ISBN 90-277-0704-9; Pb 90-277-0705-7
51. K.H. Wolff: *Surrender and Cath*. Experience and Inquiry Today. [Synthese Library 105] 1976   ISBN 90-277-0758-8; Pb 90-277-0765-0
52. K. Kosík: *Dialectics of the Concrete*. A Study on Problems of Man and World. 1976
ISBN 90-277-0761-8; Pb 90-277-0764-2
53. N. Goodman: *The Structure of Appearance*. [Synthese Library 107] 1977
ISBN 90-277-0773-1; Pb 90-277-0774-X
54. H.A. Simon: *Models of Discovery* and Other Topics in the Methods of Science. [Synthese Library 114] 1977   ISBN 90-277-0812-6; Pb 90-277-0858-4
55. M. Lazerowitz: *The Language of Philosophy*. Freud and Wittgenstein. [Synthese Library 117] 1977   ISBN 90-277-0826-6; Pb 90-277-0862-2
56. T. Nickles (ed.): *Scientific Discovery, Logic, and Rationality*. 1980
ISBN 90-277-1069-4; Pb 90-277-1070-8

Boston Studies in the Philosophy of Science

57. J. Margolis: *Persons and Mind.* The Prospects of Nonreductive Materialism. [Synthese Library 121] 1978   ISBN 90-277-0854-1; Pb 90-277-0863-0
58. G. Radnitzky and G. Andersson (eds.): *Progress and Rationality in Science.* [Synthese Library 125] 1978   ISBN 90-277-0921-1; Pb 90-277-0922-X
59. G. Radnitzky and G. Andersson (eds.): *The Structure and Development of Science.* [Synthese Library 136] 1979   ISBN 90-277-0994-7; Pb 90-277-0995-5
60. T. Nickles (ed.): *Scientific Discovery.* Case Studies. 1980   ISBN 90-277-1092-9; Pb 90-277-1093-7
61. M.A. Finocchiaro: *Galileo and the Art of Reasoning.* Rhetorical Foundation of Logic and Scientific Method. 1980   ISBN 90-277-1094-5; Pb 90-277-1095-3
62. W.A. Wallace: *Prelude to Galileo.* Essays on Medieval and 16th-Century Sources of Galileo's Thought. 1981   ISBN 90-277-1215-8; Pb 90-277-1216-6
63. F. Rapp: *Analytical Philosophy of Technology.* Translated from German. 1981   ISBN 90-277-1221-2; Pb 90-277-1222-0
64. R.S. Cohen and M.W. Wartofsky (eds.): *Hegel and the Sciences.* 1984   ISBN 90-277-0726-X
65. J. Agassi: *Science and Society.* Studies in the Sociology of Science. 1981   ISBN 90-277-1244-1; Pb 90-277-1245-X
66. L. Tondl: *Problems of Semantics.* A Contribution to the Analysis of the Language of Science. Translated from Czech. 1981   ISBN 90-277-0148-2; Pb 90-277-0316-7
67. J. Agassi and R.S. Cohen (eds.): *Scientific Philosophy Today.* Essays in Honor of Mario Bunge. 1982   ISBN 90-277-1262-X; Pb 90-277-1263-8
68. W. Krajewski (ed.): *Polish Essays in the Philosophy of the Natural Sciences.* Translated from Polish and edited by R.S. Cohen and C.R. Fawcett. 1982   ISBN 90-277-1286-7; Pb 90-277-1287-5
69. J.H. Fetzer: *Scientific Knowledge.* Causation, Explanation and Corroboration. 1981   ISBN 90-277-1335-9; Pb 90-277-1336-7
70. S. Grossberg: *Studies of Mind and Brain.* Neural Principles of Learning, Perception, Development, Cognition, and Motor Control. 1982   ISBN 90-277-1359-6; Pb 90-277-1360-X
71. R.S. Cohen and M.W. Wartofsky (eds.): *Epistemology, Methodology, and the Social Sciences.* 1983.   ISBN 90-277-1454-1
72. K. Berka: *Measurement.* Its Concepts, Theories and Problems. Translated from Czech. 1983   ISBN 90-277-1416-9
73. G.L. Pandit: *The Structure and Growth of Scientific Knowledge.* A Study in the Methodology of Epistemic Appraisal. 1983   ISBN 90-277-1434-7
74. A.A. Zinov'ev: *Logical Physics.* Translated from Russian. Edited by R.S. Cohen. 1983 [*see also* Volume 9]   ISBN 90-277-0734-0
75. G-G. Granger: *Formal Thought and the Sciences of Man.* Translated from French. With and Introduction by A. Rosenberg. 1983   ISBN 90-277-1524-6
76. R.S. Cohen and L. Laudan (eds.): *Physics, Philosophy and Psychoanalysis.* Essays in Honor of Adolf Grünbaum. 1983   ISBN 90-277-1533-5
77. G. Böhme, W. van den Daele, R. Hohlfeld, W. Krohn and W. Schäfer: *Finalization in Science.* The Social Orientation of Scientific Progress. Translated from German. Edited by W. Schäfer. 1983   ISBN 90-277-1549-1
78. D. Shapere: *Reason and the Search for Knowledge.* Investigations in the Philosophy of Science. 1984   ISBN 90-277-1551-3; Pb 90-277-1641-2
79. G. Andersson (ed.): *Rationality in Science and Politics.* Translated from German. 1984   ISBN 90-277-1575-0; Pb 90-277-1953-5

Boston Studies in the Philosophy of Science

80. P.T. Durbin and F. Rapp (eds.): *Philosophy and Technology*. [*Also* Philosophy and Technology Series, Vol. 1] 1983　ISBN 90-277-1576-9
81. M. Marković: *Dialectical Theory of Meaning*. Translated from Serbo-Croat. 1984
　ISBN 90-277-1596-3
82. R.S. Cohen and M.W. Wartofsky (eds.): *Physical Sciences and History of Physics*. 1984　ISBN 90-277-1615-3
83. É. Meyerson: *The Relativistic Deduction*. Epistemological Implications of the Theory of Relativity. Translated from French. With a Review by Albert Einstein and an Introduction by Milič Čapek. 1985　ISBN 90-277-1699-4
84. R.S. Cohen and M.W. Wartofsky (eds.): *Methodology, Metaphysics and the History of Science*. In Memory of Benjamin Nelson. 1984　ISBN 90-277-1711-7
85. G. Tamás: *The Logic of Categories*. Translated from Hungarian. Edited by R.S. Cohen. 1986　ISBN 90-277-1742-7
86. S.L. de C. Fernandes: *Foundations of Objective Knowledge*. The Relations of Popper's Theory of Knowledge to That of Kant. 1985　ISBN 90-277-1809-1
87. R.S. Cohen and T. Schnelle (eds.): *Cognition and Fact*. Materials on Ludwik Fleck. 1986　ISBN 90-277-1902-0
88. G. Freudenthal: *Atom and Individual in the Age of Newton*. On the Genesis of the Mechanistic World View. Translated from German. 1986　ISBN 90-277-1905-5
89. A. Donagan, A.N. Perovich Jr and M.V. Wedin (eds.): *Human Nature and Natural Knowledge*. Essays presented to Marjorie Grene on the Occasion of Her 75th Birthday. 1986　ISBN 90-277-1974-8
90. C. Mitcham and A. Hunning (eds.): *Philosophy and Technology II*. Information Technology and Computers in Theory and Practice. [*Also* Philosophy and Technology Series, Vol. 2] 1986　ISBN 90-277-1975-6
91. M. Grene and D. Nails (eds.): *Spinoza and the Sciences*. 1986
　ISBN 90-277-1976-4
92. S.P. Turner: *The Search for a Methodology of Social Science*. Durkheim, Weber, and the 19th-Century Problem of Cause, Probability, and Action. 1986.
　ISBN 90-277-2067-3
93. I.C. Jarvie: *Thinking about Society*. Theory and Practice. 1986
　ISBN 90-277-2068-1
94. E. Ullmann-Margalit (ed.): *The Kaleidoscope of Science*. The Israel Colloquium: Studies in History, Philosophy, and Sociology of Science, Vol. 1. 1986
　ISBN 90-277-2158-0; Pb 90-277-2159-9
95. E. Ullmann-Margalit (ed.): *The Prism of Science*. The Israel Colloquium: Studies in History, Philosophy, and Sociology of Science, Vol. 2. 1986
　ISBN 90-277-2160-2; Pb 90-277-2161-0
96. G. Márkus: *Language and Production*. A Critique of the Paradigms. Translated from French. 1986　ISBN 90-277-2169-6
97. F. Amrine, F.J. Zucker and H. Wheeler (eds.): *Goethe and the Sciences: A Reappraisal*. 1987　ISBN 90-277-2265-X; Pb 90-277-2400-8
98. J.C. Pitt and M. Pera (eds.): *Rational Changes in Science*. Essays on Scientific Reasoning. Translated from Italian. 1987　ISBN 90-277-2417-2
99. O. Costa de Beauregard: *Time, the Physical Magnitude*. 1987
　ISBN 90-277-2444-X
100. A. Shimony and D. Nails (eds.): *Naturalistic Epistemology*. A Symposium of Two Decades. 1987　ISBN 90-277-2337-0

Boston Studies in the Philosophy of Science

101. N. Rotenstreich: *Time and Meaning in History.* 1987   ISBN 90-277-2467-9
102. D.B. Zilberman: *The Birth of Meaning in Hindu Thought.* Edited by R.S. Cohen. 1988   ISBN 90-277-2497-0
103. T.F. Glick (ed.): *The Comparative Reception of Relativity.* 1987
      ISBN 90-277-2498-9
104. Z. Harris, M. Gottfried, T. Ryckman, P. Mattick Jr, A. Daladier, T.N. Harris and S. Harris: *The Form of Information in Science.* Analysis of an Immunology Sublanguage. With a Preface by Hilary Putnam. 1989   ISBN 90-277-2516-0
105. F. Burwick (ed.): *Approaches to Organic Form.* Permutations in Science and Culture. 1987   ISBN 90-277-2541-1
106. M. Almási: *The Philosophy of Appearances.* Translated from Hungarian. 1989
      ISBN 90-277-2150-5
107. S. Hook, W.L. O'Neill and R. O'Toole (eds.): *Philosophy, History and Social Action.* Essays in Honor of Lewis Feuer. With an Autobiographical Essay by L. Feuer. 1988
      ISBN 90-277-2644-2
108. I. Hronszky, M. Fehér and B. Dajka: *Scientific Knowledge Socialized.* Selected Proceedings of the 5th Joint International Conference on the History and Philosophy of Science organized by the IUHPS (Veszprém, Hungary, 1984). 1988
      ISBN 90-277-2284-6
109. P. Tillers and E.D. Green (eds.): *Probability and Inference in the Law of Evidence.* The Uses and Limits of Bayesianism. 1988   ISBN 90-277-2689-2
110. E. Ullmann-Margalit (ed.): *Science in Reflection.* The Israel Colloquium: Studies in History, Philosophy, and Sociology of Science, Vol. 3. 1988
      ISBN 90-277-2712-0; Pb 90-277-2713-9
111. K. Gavroglu, Y. Goudaroulis and P. Nicolacopoulos (eds.): *Imre Lakatos and Theories of Scientific Change.* 1989   ISBN 90-277-2766-X
112. B. Glassner and J.D. Moreno (eds.): *The Qualitative-Quantitative Distinction in the Social Sciences.* 1989   ISBN 90-277-2829-1
113. K. Arens: *Structures of Knowing.* Psychologies of the 19th Century. 1989
      ISBN 0-7923-0009-2
114. A. Janik: *Style, Politics and the Future of Philosophy.* 1989   ISBN 0-7923-0056-4
115. F. Amrine (ed.): *Literature and Science as Modes of Expression.* With an Introduction by S. Weininger. 1989   ISBN 0-7923-0133-1
116. J.R. Brown and J. Mittelstrass (eds.): *An Intimate Relation.* Studies in the History and Philosophy of Science. Presented to Robert E. Butts on His 60th Birthday. 1989
      ISBN 0-7923-0169-2
117. F. D'Agostino and I.C. Jarvie (eds.): *Freedom and Rationality.* Essays in Honor of John Watkins. 1989   ISBN 0-7923-0264-8
118. D. Zolo: *Reflexive Epistemology.* The Philosophical Legacy of Otto Neurath. 1989
      ISBN 0-7923-0320-2
119. M. Kearn, B.S. Philips and R.S. Cohen (eds.): *Georg Simmel and Contemporary Sociology.* 1989   ISBN 0-7923-0407-1
120. T.H. Levere and W.R. Shea (eds.): *Nature, Experiment and the Science.* Essays on Galileo and the Nature of Science. In Honour of Stillman Drake. 1989
      ISBN 0-7923-0420-9
121. P. Nicolacopoulos (ed.): *Greek Studies in the Philosophy and History of Science.* 1990   ISBN 0-7923-0717-8

Boston Studies in the Philosophy of Science

122. R. Cooke and D. Costantini (eds.): *Statistics in Science*. The Foundations of Statistical Methods in Biology, Physics and Economics. 1990   ISBN 0-7923-0797-6
123. P. Duhem: *The Origins of Statics*. Translated from French by G.F. Leneaux, V.N. Vagliente and G.H. Wagner. With an Introduction by S.L. Jaki. 1991
 ISBN 0-7923-0898-0
124. H. Kamerlingh Onnes: *Through Measurement to Knowledge*. The Selected Papers, 1853–1926. Edited and with an Introduction by K. Gavroglu and Y. Goudaroulis. 1991   ISBN 0-7923-0825-5
125. M. Čapek: *The New Aspects of Time: Its Continuity and Novelties*. Selected Papers in the Philosophy of Science. 1991   ISBN 0-7923-0911-1
126. S. Unguru (ed.): *Physics, Cosmology and Astronomy, 1300–1700*. Tension and Accommodation. 1991   ISBN 0-7923-1022-5
127. Z. Bechler: *Newton's Physics on the Conceptual Structure of the Scientific Revolution*. 1991   ISBN 0-7923-1054-3
128. É. Meyerson: *Explanation in the Sciences*. Translated from French by M-A. Siple and D.A. Siple. 1991   ISBN 0-7923-1129-9
129. A.I. Tauber (ed.): *Organism and the Origins of Self*. 1991   ISBN 0-7923-1185-X
130. F.J. Varela and J-P. Dupuy (eds.): *Understanding Origins*. Contemporary Views on the Origin of Life, Mind and Society. 1992   ISBN 0-7923-1251-1
131. G.L. Pandit: *Methodological Variance*. Essays in Epistemological Ontology and the Methodology of Science. 1991   ISBN 0-7923-1263-5
132. G. Munévar (ed.): *Beyond Reason*. Essays on the Philosophy of Paul Feyerabend. 1991   ISBN 0-7923-1272-4
133. T.E. Uebel (ed.): *Rediscovering the Forgotten Vienna Circle*. Austrian Studies on Otto Neurath and the Vienna Circle. Partly translated from German. 1991
 ISBN 0-7923-1276-7
134. W.R. Woodward and R.S. Cohen (eds.): *World Views and Scientific Discipline Formation*. Science Studies in the [former] German Democratic Republic. Partly translated from German by W.R. Woodward. 1991   ISBN 0-7923-1286-4
135. P. Zambelli: *The Speculum Astronomiae and Its Enigma*. Astrology, Theology and Science in Albertus Magnus and His Contemporaries. 1992   ISBN 0-7923-1380-1
136. P. Petitjean, C. Jami and A.M. Moulin (eds.): *Science and Empires*. Historical Studies about Scientific Development and European Expansion.   ISBN 0-7923-1518-9
137. W.A. Wallace: *Galileo's Logic of Discovery and Proof*. The Background, Content, and Use of His Appropriated Treatises on Aristotle's *Posterior Analytics*. 1992
 ISBN 0-7923-1577-4
138. W.A. Wallace: *Galileo's Logical Treatises*. A Translation, with Notes and Commentary, of His Appropriated Latin Questions on Aristotle's *Posterior Analytics*. 1992   ISBN 0-7923-1578-2
 Set (137 + 138) ISBN 0-7923-1579-0
139. M.J. Nye, J.L. Richards and R.H. Stuewer (eds.): *The Invention of Physical Science*. Intersections of Mathematics, Theology and Natural Philosophy since the Seventeenth Century. Essays in Honor of Erwin N. Hiebert. 1992
 ISBN 0-7923-1753-X
140. G. Corsi, M.L. dalla Chiara and G.C. Ghirardi (eds.): *Bridging the Gap: Philosophy, Mathematics and Physics*. Lectures on the Foundations of Science. 1992
 ISBN 0-7923-1761-0

Boston Studies in the Philosophy of Science

141. C.-H. Lin and D. Fu (eds.): *Philosophy and Conceptual History of Science in Taiwan.* 1992 ISBN 0-7923-1766-1
142. S. Sarkar (ed.): *The Founders of Evolutionary Genetics.* A Centenary Reappraisal. 1992 ISBN 0-7923-1777-7
143. J. Blackmore (ed.): *Ernst Mach – A Deeper Look.* Documents and New Perspectives. 1992 ISBN 0-7923-1853-6
144. P. Kroes and M. Bakker (eds.): *Technological Development and Science in the Industrial Age.* New Perspectives on the Science–Technology Relationship. 1992 ISBN 0-7923-1898-6
145. S. Amsterdamski: *Between History and Method.* Disputes about the Rationality of Science. 1992 ISBN 0-7923-1941-9
146. E. Ullmann-Margalit (ed.): *The Scientific Enterprise.* The Bar-Hillel Colloquium: Studies in History, Philosophy, and Sociology of Science, Volume 4. 1992 ISBN 0-7923-1992-3
147. L. Embree (ed.): *Metaarchaeology.* Reflections by Archaeologists and Philosophers. 1992 ISBN 0-7923-2023-9
148. S. French and H. Kamminga (eds.): *Correspondence, Invariance and Heuristics.* Essays in Honour of Heinz Post. 1993 ISBN 0-7923-2085-9
149. M. Bunzl: *The Context of Explanation.* 1993 ISBN 0-7923-2153-7
150. I.B. Cohen (ed.): *The Natural Sciences and the Social Sciences.* Some Critical and Historical Perspectives. 1994 ISBN 0-7923-2223-1
151. K. Gavroglu, Y. Christianidis and E. Nicolaidis (eds.): *Trends in the Historiography of Science.* 1994 ISBN 0-7923-2255-X
152. S. Poggi and M. Bossi (eds.): *Romanticism in Science.* Science in Europe, 1790–1840. 1994 ISBN 0-7923-2336-X
153. J. Faye and H.J. Folse (eds.): *Niels Bohr and Contemporary Philosophy.* 1994 ISBN 0-7923-2378-5
154. C.C. Gould and R.S. Cohen (eds.): *Artifacts, Representations, and Social Practice.* Essays for Marx W. Wartofsky. 1994 ISBN 0-7923-2481-1
155. R.E. Butts: *Historical Pragmatics.* Philosophical Essays. 1993 ISBN 0-7923-2498-6
156. R. Rashed: *The Development of Arabic Mathematics: Between Arithmetic and Algebra.* Translated from French by A.F.W. Armstrong. 1994 ISBN 0-7923-2565-6
157. I. Szumilewicz-Lachman (ed.): *Zygmunt Zawirski: His Life and Work.* With Selected Writings on Time, Logic and the Methodology of Science. Translations by Feliks Lachman. Ed. by R.S. Cohen, with the assistance of B. Bergo. 1994 ISBN 0-7923-2566-4
158. S.N. Haq: *Names, Natures and Things.* The Alchemist Jabir ibn Ḥanyyān and His *Kitāb al-Ahjār* (Book of Stones). 1994 ISBN 0-7923-2587-7
159. P. Plaass: *Kant's Theory of Natural Science.* Translation, Analytic Introduction and Commentary by Alfred E. and Maria G. Miller. 1994 ISBN 0-7923-2750-0
160. J. Misiek (ed.): *The Problem of Rationality in Science and its Philosophy.* On Popper vs. Polanyi. The Polish Conferences 1988–89. 1995 ISBN 0-7923-2925-2
161. I.C. Jarvie and N. Laor (eds.): *Critical Rationalism, Metaphysics and Science.* Essays for Joseph Agassi, Volume I. 1995 ISBN 0-7923-2960-0

Boston Studies in the Philosophy of Science

162. I.C. Jarvie and N. Laor (eds.): *Critical Rationalism, the Social Sciences and the Humanities.* Essays for Joseph Agassi, Volume II. 1995 ISBN 0-7923-2961-9
    Set (161–162) ISBN 0-7923-2962-7
163. K. Gavroglu, J. Stachel and M.W. Wartofsky (eds.): *Physics, Philosophy, and the Scientific Community.* Essays in the Philosophy and History of the Natural Sciences and Mathematics. In Honor of Robert S. Cohen. 1995 ISBN 0-7923-2988-0
164. K. Gavroglu, J. Stachel and M.W. Wartofsky (eds.): *Science, Politics and Social Practice.* Essays on Marxism and Science, Philosophy of Culture and the Social Sciences. In Honor of Robert S. Cohen. 1995 ISBN 0-7923-2989-9
165. K. Gavroglu, J. Stachel and M.W. Wartofsky (eds.): *Science, Mind and Art.* Essays on Science and the Humanistic Understanding in Art, Epistemology, Religion and Ethics. Essays in Honor of Robert S. Cohen. 1995 ISBN 0-7923-2990-2
    Set (163–165) ISBN 0-7923-2991-0
166. K.H. Wolff: *Transformation in the Writing.* A Case of Surrender-and-Catch. 1995
    ISBN 0-7923-3178-8
167. A.J. Kox and D.M. Siegel (eds.): *No Truth Except in the Details.* Essays in Honor of Martin J. Klein. 1995 ISBN 0-7923-3195-8
168. J. Blackmore: *Ludwig Boltzmann, His Later Life and Philosophy, 1900–1906.* Book One: A Documentary History. 1995 ISBN 0-7923-3231-8
169. R.S. Cohen, R. Hilpinen and R. Qiu (eds.): *Realism and Anti-Realism in the Philosophy of Science.* Beijing International Conference, 1992. 1996
    ISBN 0-7923-3233-4
170. I. Kuçuradi and R.S. Cohen (eds.): *The Concept of Knowledge.* The Ankara Seminar. 1995 ISBN 0-7923-3241-5
171. M.A. Grodin (ed.): *Meta Medical Ethics*: The Philosophical Foundations of Bioethics. 1995 ISBN 0-7923-3344-6
172. S. Ramirez and R.S. Cohen (eds.): *Mexican Studies in the History and Philosophy of Science.* 1995 ISBN 0-7923-3462-0
173. C. Dilworth: *The Metaphysics of Science.* An Account of Modern Science in Terms of Principles, Laws and Theories. 1995 ISBN 0-7923-3693-3
174. J. Blackmore: *Ludwig Boltzmann, His Later Life and Philosophy, 1900–1906* Book Two: The Philosopher. 1995 ISBN 0-7923-3464-7
175. P. Damerow: *Abstraction and Representation.* Essays on the Cultural Evolution of Thinking. 1996 ISBN 0-7923-3816-2
176. M.S. Macrakis: *Scarcity's Ways: The Origins of Capital.* A Critical Essay on Thermodynamics, Statistical Mechanics and Economics. 1997
    ISBN 0-7923-4760-9
177. M. Marion and R.S. Cohen (eds.): *Québec Studies in the Philosophy of Science.* Part I: Logic, Mathematics, Physics and History of Science. Essays in Honor of Hugues Leblanc. 1995 ISBN 0-7923-3559-7
178. M. Marion and R.S. Cohen (eds.): *Québec Studies in the Philosophy of Science.* Part II: Biology, Psychology, Cognitive Science and Economics. Essays in Honor of Hugues Leblanc. 1996 ISBN 0-7923-3560-0
    Set (177–178) ISBN 0-7923-3561-9
179. Fan Dainian and R.S. Cohen (eds.): *Chinese Studies in the History and Philosophy of Science and Technology.* 1996 ISBN 0-7923-3463-9

Boston Studies in the Philosophy of Science

180. P. Forman and J.M. Sánchez-Ron (eds.): *National Military Establishments and the Advancement of Science and Technology.* Studies in 20th Century History. 1996
    ISBN 0-7923-3541-4
181. E.J. Post: *Quantum Reprogramming.* Ensembles and Single Systems: A Two-Tier Approach to Quantum Mechanics. 1995 ISBN 0-7923-3565-1
182. A.I. Tauber (ed.): *The Elusive Synthesis: Aesthetics and Science.* 1996
    ISBN 0-7923-3904-5
183. S. Sarkar (ed.): *The Philosophy and History of Molecular Biology: New Perspectives.* 1996 ISBN 0-7923-3947-9
184. J.T. Cushing, A. Fine and S. Goldstein (eds.): *Bohmian Mechanics and Quantum Theory: An Appraisal.* 1996 ISBN 0-7923-4028-0
185. K. Michalski: *Logic and Time.* An Essay on Husserl's Theory of Meaning. 1996
    ISBN 0-7923-4082-5
186. G. Munévar (ed.): *Spanish Studies in the Philosophy of Science.* 1996
    ISBN 0-7923-4147-3
187. G. Schubring (ed.): *Hermann Günther Graßmann (1809–1877): Visionary Mathematician, Scientist and Neohumanist Scholar.* Papers from a Sesquicentennial Conference. 1996 ISBN 0-7923-4261-5
188. M. Bitbol: *Schrödinger's Philosophy of Quantum Mechanics.* 1996
    ISBN 0-7923-4266-6
189. J. Faye, U. Scheffler and M. Urchs (eds.): *Perspectives on Time.* 1997
    ISBN 0-7923-4330-1
190. K. Lehrer and J.C. Marek (eds.): *Austrian Philosophy Past and Present.* Essays in Honor of Rudolf Haller. 1996 ISBN 0-7923-4347-6
191. J.L. Lagrange: *Analytical Mechanics.* Translated and edited by Auguste Boissonade and Victor N. Vagliente. Translated from the *Mécanique Analytique, novelle édition* of 1811. 1997 ISBN 0-7923-4349-2
192. D. Ginev and R.S. Cohen (eds.): *Issues and Images in the Philosophy of Science.* Scientific and Philosophical Essays in Honour of Azarya Polikarov. 1997
    ISBN 0-7923-4444-8
193. R.S. Cohen, M. Horne and J. Stachel (eds.): *Experimental Metaphysics.* Quantum Mechanical Studies for Abner Shimony, Volume One. 1997 ISBN 0-7923-4452-9
194. R.S. Cohen, M. Horne and J. Stachel (eds.): *Potentiality, Entanglement and Passion-at-a-Distance.* Quantum Mechanical Studies for Abner Shimony, Volume Two. 1997
    ISBN 0-7923-4453-7; Set 0-7923-4454-5
195. R.S. Cohen and A.I. Tauber (eds.): *Philosophies of Nature: The Human Dimension.* 1997 ISBN 0-7923-4579-7
196. M. Otte and M. Panza (eds.): *Analysis and Synthesis in Mathematics.* History and Philosophy. 1997 ISBN 0-7923-4570-3
197. A. Denkel: *The Natural Background of Meaning.* 1999 ISBN 0-7923-5331-5
198. D. Baird, R.I.G. Hughes and A. Nordmann (eds.): *Heinrich Hertz: Classical Physicist, Modern Philosopher.* 1999 ISBN 0-7923-4653-X
199. A. Franklin: *Can That be Right?* Essays on Experiment, Evidence, and Science. 1999 ISBN 0-7923-5464-8
200. D. Raven, W. Krohn and R.S. Cohen (eds.): *The Social Origins of Modern Science.* 2000 ISBN 0-7923-6457-0
201. Reserved
202. Reserved

Boston Studies in the Philosophy of Science

203. B. Babich and R.S. Cohen (eds.): *Nietzsche, Theories of Knowledge, and Critical Theory*. Nietzsche and the Sciences I. 1999 ISBN 0-7923-5742-6
204. B. Babich and R.S. Cohen (eds.): *Nietzsche, Epistemology, and Philosophy of Science*. Nietzsche and the Science II. 1999 ISBN 0-7923-5743-4
205. R. Hooykaas: *Fact, Faith and Fiction in the Development of Science*. The Gifford Lectures given in the University of St Andrews 1976. 1999 ISBN 0-7923-5774-4
206. M. Fehér, O. Kiss and L. Ropolyi (eds.): *Hermeneutics and Science*. 1999 ISBN 0-7923-5798-1
207. R.M. MacLeod (ed.): *Science and the Pacific War*. Science and Survival in the Pacific, 1939–1945. 1999 ISBN 0-7923-5851-1
208. I. Hanzel: *The Concept of Scientific Law in the Philosophy of Science and Epistemology*. A Study of Theoretical Reason. 1999 ISBN 0-7923-5852-X
209. G. Helm; R.J. Deltete (ed./transl.): *The Historical Development of Energetics*. 1999 ISBN 0-7923-5874-0
210. A. Orenstein and P. Kotatko (eds.): *Knowledge, Language and Logic*. Questions for Quine. 1999 ISBN 0-7923-5986-0
211. R.S. Cohen and H. Levine (eds.): *Maimonides and the Sciences*. 2000 ISBN 0-7923-6053-2
212. H. Gourko, D.I. Williamson and A.I. Tauber (eds.): *The Evolutionary Biology Papers of Elie Metchnikoff.* 2000 ISBN 0-7923-6067-2
213. S. D'Agostino: *A History of the Ideas of Theoretical Physics*. Essays on the Nineteenth and Twentieth Century Physics. 2000 ISBN 0-7923-6094-X
214. S. Lelas: *Science and Modernity*. Toward An Integral Theory of Science. 2000 ISBN 0-7923-6303-5
215. E. Agazzi and M. Pauri (eds.): *The Reality of the Unobservable*. Observability, Unobservability and Their Impact on the Issue of Scientific Realism. 2000 ISBN 0-7923-6311-6
216. P. Hoyningen-Huene and H. Sankey (eds.): *Incommensurability and Related Matters*. 2001 ISBN 0-7923-6989-0
217. A. Nieto-Galan: *Colouring Textiles*. A History of Natural Dyestuffs in Industrial Europe. 2001 ISBN 0-7923-7022-8
218. J. Blackmore, R. Itagaki and S. Tanaka (eds.): *Ernst Mach's Vienna 1895–1930*. Or Phenomenalism as Philosophy of Science. 2001 ISBN 0-7923-7122-4
219. R. Vihalemm (ed.): *Estonian Studies in the History and Philosophy of Science*. 2001 ISBN 0-7923-7189-5
220. W. Lefèvre(ed.): *Between Leibniz, Newton, and Kant*. Philosophy and Science in the Eighteenth Century. 2001 ISBN 0-7923-7198-4
221. T.F. Glick, M.Á. Puig-Samper and R. Ruiz (eds.): *The Reception of Darwinism in the Iberian World*. Spain, Spanish America and Brazil. 2001 ISBN 1-4020-0082-0
222. U. Klein (ed.): *Tools and Modes of Representation in the Laboratory Sciences*. 2001 ISBN 1-4020-0100-2
223. P. Duhem: *Mixture and Chemical Combination*. And Related Essays. Edited and translated, with an introduction, by Paul Needham. 2002 ISBN 1-4020-0232-7
224. J.C. Boudri: *What was Mechanical about Mechanics*. The Concept of Force Between Metaphysics and Mechanics from Newton to Lagrange. 2002 ISBN 1-4020-0233-5
225. B.E. Babich (ed.): *Hermeneutic Philosophy of Science, Van Gogh's Eyes, and God*. Essays in Honor of Patrick A. Heelan, S.J. 2002 ISBN 1-4020-0234-3

Boston Studies in the Philosophy of Science

226. D. Davies Villemaire: *E.A. Burtt, Historian and Philosopher*. A Study of the Author of The Metaphysical Foundations of Modern Physical Science. 2002
 ISBN 1-4020-0428-1
227. L.J. Cohen: *Knowledge and Language*. Selected Essays of L. Jonathan Cohen. Edited and with an introduction by James Logue. 2002   ISBN 1-4020-0474-5
228. G.E. Allen and R.M. MacLeod (eds.): *Science, History and Social Activism: A Tribute to Everett Mendelsohn*. 2002   ISBN 1-4020-0495-0
229. O. Gal: *Meanest Foundations and Nobler Superstructures*. Hooke, Newton and the "Compounding of the Celestiall Motions of the Planetts". 2002
 ISBN 1-4020-0732-9
230. R. Nola: *Rescuing Reason*. A Critique of Anti-Rationalist Views of Science and Knowledge. 2003   Hb: ISBN 1-4020-1042-7; Pb ISBN 1-4020-1043-5
231. J. Agassi: *Science and Culture*. 2003   ISBN 1-4020-1156-3
232. M.C. Galavotti (ed.): *Observation and Experiment in the Natural and Social Science*. 2003   ISBN 1-4020-1251-9
233. A. Simões, A. Carneiro and M.P. Diogo (eds.): *Travels of Learning. A Geography of Science in Europe*. 2003   ISBN 1-4020-1259-4
234. A. Ashtekar, R. Cohen, D. Howard, J. Renn, S. Sarkar and A. Shimony (eds.): *Revisiting the Foundations of Relativistic Physics*. Festschrift in Honor of John Stachel. 2003   ISBN 1-4020-1284-5
235. R.P. Farell: *Feyerabend and Scientific Values*. Tightrope-Walking Rationality. 2003
 ISBN 1-4020-1350-7
236. D. Ginev (ed.): *Bulgarian Studies in the Philosophy of Science*. 2003
 ISBN 1-4020-1496-1
237. C. Sasaki: *Descartes Mathematical Thought*. 2003   ISBN 1-4020-1746-4
238. K. Chemla (ed.): *History of Science, History of Text*. 2004   ISBN 1-4020-2320-0
239. C.R. Palmerino and J.M.M.H. Thijssen (eds.): *The Reception of the Galilean Science of Motion in Seventeenth-Century Europe*. 2004   ISBN 1-4020-2454-1
240. J. Christianidis (ed.): *Classics in the History of Greek Mathematics*. 2004
 ISBN 1-4020-0081-2
241. R.M. Brain and O. Knudsen (eds.): *Hans Christian Ørsted and the Romantic Quest for Unity*. Ideas, Disciplines, Practices. 2005   ISBN 1-4020-2979-9
242. D. Baird, E. Scerri and L. McIntyre (eds.): *Philosophy of Chemistry*. Synthesis of a New Discipline. 2005   ISBN 1-4020-3256-0
243. D.B. Zilberman, H. Gourko and R.S. Cohen (eds.): *Analogy in Indian and Western Philosophical Thought*. 2005   ISBN 1-4020-3339-7
244. G. Irzik and G. Güzeldere (eds.): *Turkish Studies in the History and Philosophy of Science*. 2005   ISBN 1-4020-3332-X
245. H.E. Gruber and K. Bödeker (eds.): *Creativity, Psychology and the History of Science*. 2005   ISBN 1-4020-3491-1
246. S. Katzir: *The Beginnings of Piezoelectricity*. A Study in Mundane Physics. 2006
 ISBN 1-4020-4669-3
247. D. Ginev: *The Context of Constitution*. Beyond the Edge of Epistemological Justification. 2006   ISBN 1-4020-4712-6
248. J. Renn and K. Gavroglu (eds.): *Positioning the History of Science*. 2007
 ISBN 1-4020-5419-X

Boston Studies in the Philosophy of Science

249. M. Schemmel: *The English Galileo.* Thomas Harriot's Work on Motion as an Example of Preclassical Mechanics. 2008   ISBN 1-4020-5498-X
250. J. Renn (ed.): *The Genesis of General Relativity.* 2007   ISBN 1-4020-3999-9
251. V.F. Hendricks, K.F. Jørgensen, J. Lützen and S.A. Pedersen (eds.): *Interactions.* Mathematics, Physics and Philosophy, 1860–1930. 2006   ISBN 1-4020-5194-8
252. J. Persson and P. Ylikoski (eds.): *Rethinking Explanation.* 2007
         ISBN 1-4020-5580-3
253. J. Agassi: *Science and History.* A Reassessment of the Historiography of science. 2008   ISBN 1-4020-5631-1
254. W.R. Laird and S. Roux (eds.): *Mechanics and Natural Philosophy Before the Scientific Revolution.* 2007   ISBN 978-1-4020-5966-7
255. L. Soler, H. Sankey and P. Hoyningen-Huene (eds.): *Rethinking Scientific Change and Theory Comparison.* Stabilities, Ruptures, Incommensurabilities? 2007
         ISBN 978-1-4020-6274-2
256. W. Spohn: *Causation, Coherence and Concepts.* 2008   ISBN 978-1-4020-5473-0
257. To be assigned
258. M. Futch: *Leibniz's Metaphysics of Time and Space.* 2008
         ISBN 978-1-4020-8236-8
259. H.-W. Schmuhl: The Kaiser-Wilhelm-Institute for Anthropology, Human Heredity and Eugenics, *Crossing Boundaries.* 1927–1945. 2008   ISBN 978-1-4020-6599-6
260. S. Heim: Plant Breeding and Agrarian Research in Kaiser-Wilhelm-Institutes, *Calories, Caoutchouc, Careers.* 1933–1945. 2008   ISBN 978-1-4020-6717-4
261. T. Bonk: *Underdetermination.* An Essay on Evidence and the Limits of Natural Knowledge. 2008   ISBN 978-1-4020-6898-0

Also of interest:
R.S. Cohen and M.W. Wartofsky (eds.): *A Portrait of Twenty-Five Years Boston Colloquia for the Philosophy of Science, 1960–1985.* 1985   ISBN Pb 90-277-1971-3
Previous volumes are still available.

springer.com